AF718039

PRACTICAL GUIDE TO MOTORS and MOTOR CONTROLLERS

Intertec Electrical Group
Overland Park, KS

Copyright © 1991 INTERTEC Publishing Corporation
All rights reserved

First Printing: May, 1991
Second Printing: July, 1993

Published by
EC&M
INTERTEC Electrical Group
9800 Metcalf Avenue
Overland Park, KS

ISBN 0-87288-457-0

Please Note: The designations "National Electrical Code", "NE Code", and "NEC", where used in this book, refer to the National Electrical Code®, which is a registered trademark of the National Fire Protection Association.

CONTENTS

PAGE

PREFACE

THIS BOOK contains practical information needed by the engineer, designer, contractor, electrician, and facility electrical maintenance and operating personnel in order to understand the applications of motors and their associated controls. Most of the material presented in this book has been edited from feature articles and departments that have appeared in the pages of *EC&M* magazine.

A great deal of attention has been paid to meet the needs of the reader by providing clear, non-jargonized text. As in the magazine itself, the articles have been written by persons who have hands-on experience as a reference for persons in the industry. Thus, only theory that is absolutely necessary for the reader's understanding of the subject is presented in the chapters of this book.

Standard motors have only one function–to rotate their shaft. The task of the designer/engineer is to properly harness this rotary motion (or linear or step motion) to a task. Horsepower, torque, full-load current, and many other parameters must be taken into consideration. But motors must also be controlled, either by starting and stopping, or by varying their speed. Determining the exact sequence of motor operation and selection of motor controllers and control elements is a major challenge.

Installation of motors seems, on the surface, a simple task. Yet, there are many techniques for properly mounting, aligning, and securing that must be properly carried out if vibration is to be prevented. Lubrication is another of the essential installation tasks. Termination of feeder cable to the motor leads, grounding, and checkout of rotation are also important tasks.

Electric motors provide the muscle needed to drive pumps and blowers, move conveyor belts and transit vehicles, and an endless amount of other tasks. Motors and their controls are rugged, long-lived items, but maintenance personnel must listen to motors as they operate, take their temperatures, and feed them with the proper voltage, in order to maintain them troublefree and dependable. The controls must be kept free of accumulated dust and dirt, checked for wear, and be allowed to dissipate heat if unexpected shutdown of systems are not to occur.

Motors and their controllers, therefore, are the concern of everyone in the electrical industry. This book is a companion to **"Understanding NE Code Rules On Motors & Motor Controls,"** also published by *EC&M*. There, the rules of the code that apply to motors and their controllers are covered in greater detail and referenced to specific code sections.

New York, NY
March 1991

Alfred Berutti
Chief Editor, EC&M

MOTOR BASICS

1

AC SQUIRREL-CAGE induction motors and DC motors have long been the workhorses of industry. Many other motor types, including synchonous, wound rotor, single phase, and specialties such as stepper motors, and others fill essential niches. None, however, have the wide application possibilities of the AC squirrel-cage induction motor and the DC motor. Thus, it is important to understand the operating principles and characteristics of these two types of motors.

THE INDUCTION MOTOR

The rotor of the AC squirrel-cage induction motor shown in **Fig. 1.1** consists of a structure of steel laminations mounted on a shaft. Embedded in the rotor is the rotor winding, which is a series of copper or aluminum bars that are all short-circuited at each end by a metallic end ring. The stator consists of steel laminations mounted in a frame. Slots in the stator hold stator windings that can be either copper or aluminum wire coils or bars. These are connected to form a circuit.

Energizing the stator coils with an AC supply voltage causes current to flow in the coils. The current produces an electromagnetic field that, in turn, causes magnetic poles to be created in the stator iron. The strength and polarity of these poles vary as the AC current flows in one direction, then the other. This change causes the poles around the stator to alternate between being south or north poles, in effect producing a rotating magnetic field.

The rotating magnetic field cuts through the rotor, inducing a current in the rotor bars. This induced current only circulates in the rotor, which in turn creates a rotor magnetic field. As

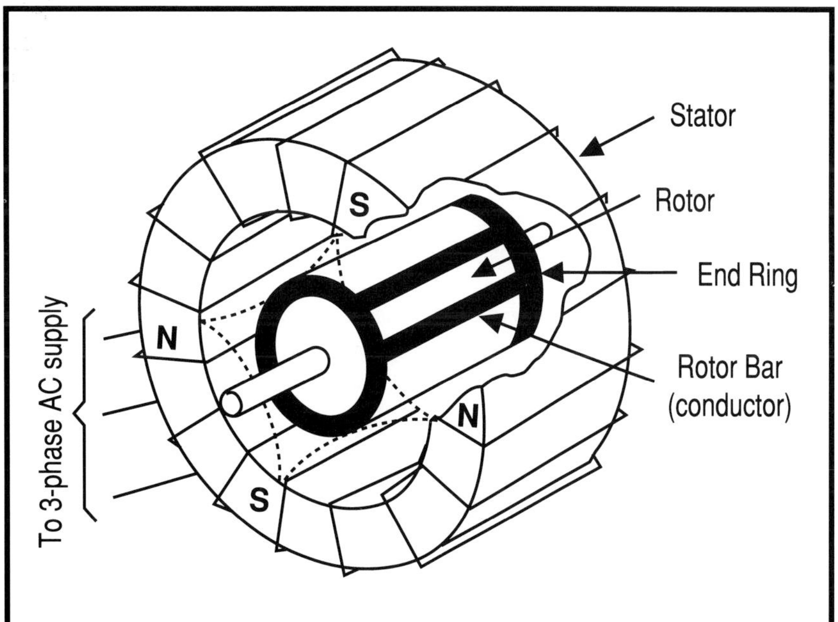

Fig. 1.1. Three-phase AC current flowing in the stator windings create a magnetic field, creating north and south poles that revolve around the stator. Magnetic forces in the rotor tend to follow the stator magnetic field, producing rotary motion.

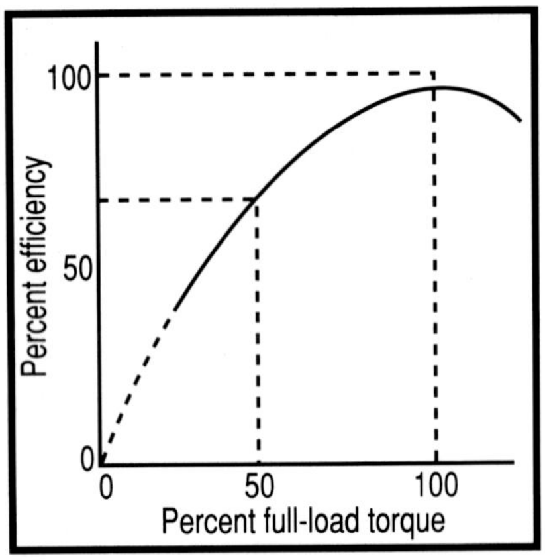

Fig. 1.2. Typical torque vs efficiency curve of a motor.

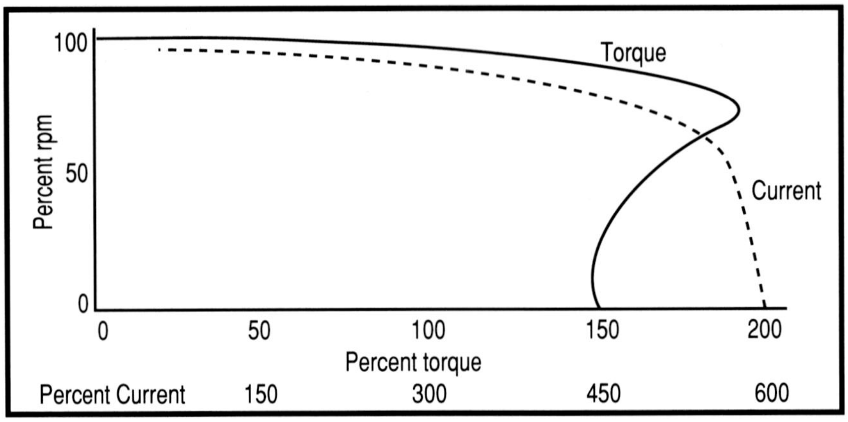

Fig. 1.3. Typical characteristic curve of rpm vs torque and current for a NEMA Design B induction motor.

with two bar magnets, the north pole of the rotor field attempts to line up with the south pole of the stator magnetic field, and the south pole to line up with the north pole. However, because the stator magnetic field is rotating, the rotor "chases" the stator field. The rotor field never quite catches up due to the need to furnish torque to the mechanical load.

INDUCTION MOTOR CHARACTERISTICS

It is the design of the motor that dictates the specific motor characteristics, such as horsepower, torque, speed, power factor, etc. A 3-phase induction motor is furnished with three windings connected to a 3-phase AC power source. When a polyphase alternating current flows in the stator winding, north and south poles are created in the stator, depending on how the windings are arranged and connected. The machine will always have at least two poles and a rotating magnetic field.

Speed. The speed (in rpm) at which an induction motor rotates is dependent on the speed of the stator rotating field and is equal to:

$$S = (120 \times f) / P$$

where f is the frequency in Hz of the source, and P is the number of poles.

For example, a motor having two poles supplied from a 60Hz source will run at 3600 rpm; a 4-pole motor will run at 1800 rpm. The actual speed of the motor is slightly less because of "slip."

Slip is the difference between the speed of the stator magnetic field and the speed of the rotor. Slip is necessary to permit the motor action to occur. Under load, the rotor slows down and the speed adjusts itelf to the point where the forces exerted by the magnetic field on the rotor is sufficient to overcome the torque requirements of the load. The resulting rotor speed is slightly less than that of the stator rotating field. For example, the actual speed of the 4-pole motor will be about 1725 rpm.

The slip necessary to carry full load depends on the motor characteristics. In general, the higher the inrush current required to start the motor, the lower the slip at which the motor can carry full load, and the higher the efficiency. The lower the inrush current required, the higher will be the slip and the lower the efficiency. The slip at rated load may vary from 3% to 20% for different types of motors.

Torque relates to the demands of the load being driven and the capability of the motor to drive that load.

- Motor torque as the turning effort exerted by the motor shaft and is commonly measured in pound-feet.
- Load torque is similarly defined as the turning effort demanded at the input shaft of the load for its proper operation. It is also commonly measured in pound-feet.

The term horsepower (hp) is more familiar than the term torque. This probably results because U.S. motor ratings and nameplates are standardized in terms of hp and speed with no mention of torque. However, hp and torque are related by the equation:

$$hp = (torque \times rpm) / 5250.$$

Fig. 1.2 is a typical torque/efficiency curve that illustrates how a motor operates most efficiently at full-load torque. Efficiency falls off rapidly as load torque decreases. The dotted line shows that at 50% of full-load torque, the efficiency at full load (93%) drops to 70%. It should be noted that typical modern motors, and particularly premium energy-efficient motors, will show higher efficiency values. In these, motor efficiency from 50% to 100% full load is relatively flat. Some motors reach their peak efficiency at 75% of full load.

A motor must be matched as closely as possible to the requirements of the driven load to obtain the most effective operation at the lowest possible operating cost. Torque demanded by the load, as well as torque available at the motor output, is the most important criteria. Motor characteristic curves, such as the one shown in **Fig. 1.3** are available for each motor to show the motor torque developed at any given motor speed.

THE EFFECT OF FREQUENCY

When a 3-phase induction motor is operated at different frequencies while the supply voltage is kept constant, the torque developed by the motor at a given speed will be approximately proportional to the square of the supply frequency. The synchronous speed of the motor will be directly proportional to the supply frequency, and the cur-

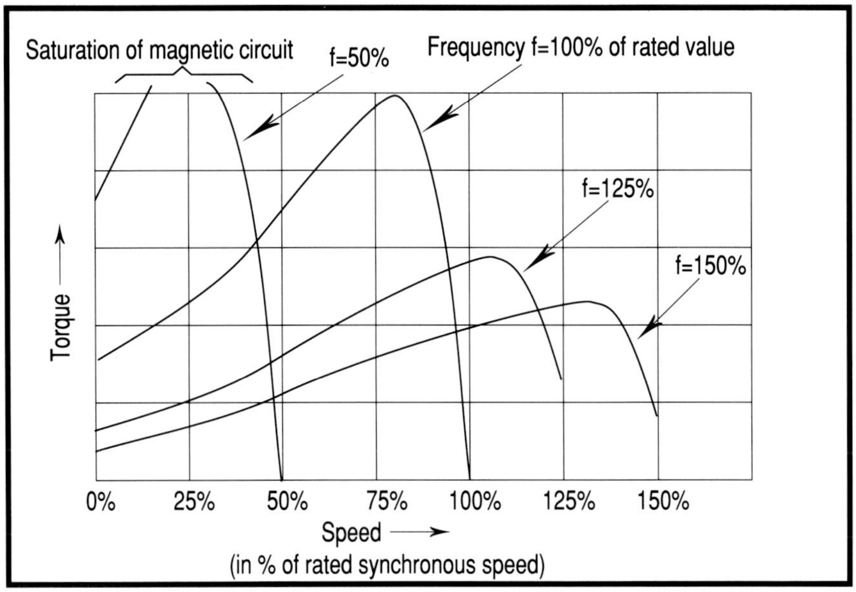

Fig. 1.4. Varying frequency while maintaining voltage constant.

rent drawn by the motor at a given speed will be (approximately) inversely proportional to the supply frequency.

A reduction in supply frequency reduces the magnetizing reactance. Therefore, if the supply voltage is kept constant, the magnetizing current (and consequently the magnetic flux in the air gap) will increase. This in turn increases the motor current. At very low frequencies, the magnetic circuit of the motor will operate far beyond saturation and the magnetizing current will become intolerably high. **Fig. 1.4** shows the typical results of varying frequency while keeping voltage constant. Thus, varying motor speed by varying frequency alone is not practical.

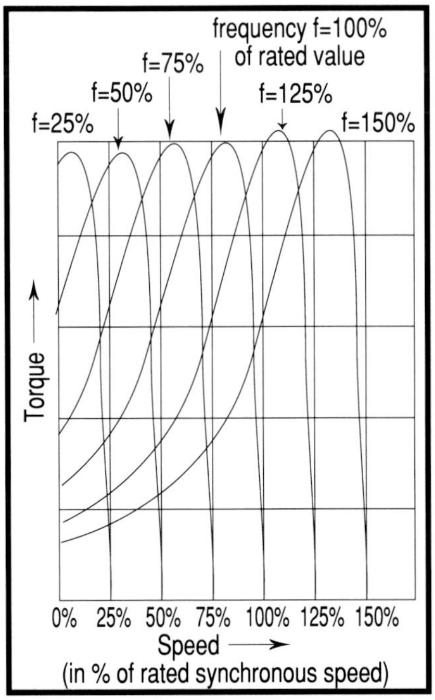

Fig. 1.5. Effects of varying both voltage and frequency while maintaining a constant ratio between both.

To maintain a constant torque, the ratio of voltage-to-frequency must be kept constant. In order to achieve this, the voltage must be varied simultaneously with the frequency. This is an excellent way to vary the speed of a squirrel-cage induction motor.

When the supply voltage and frequency are varied so that the ratio of the voltage-to-frequency remains constant, the flux density in the air gap remains almost constant. If the resistance of the stator winding is neglected, the value of the maximum torque decreases little with a decrease in frequency. The slip for maximum torque varies inversely to the frequency, and maximum torque occurs at a lower speed when the frequency is decreased. The torque/speed curves for different frequencies with constant volts/frequency ratio are shown in **Fig. 1.5.**

THE EFFECT OF VOLTAGE

AC induction motors operate better and last longer when they are applied at their rated voltage. Even though standards allow for operation at +/- 10% of rated voltage, the best operation occurs at rated voltage.

In order to drive a fixed mechanical load connected to its shaft, a motor must draw a comparable quantity of power from the line. The amount of power drawn is equal to:

P = E I (applied voltage x current).

Thus, when voltage is low, the current must become higher to provide the same amount of power as it would at rated voltage.

Low voltage on a fully loaded motor causes excessive current that overheats the motor windings. The fact that the motor current is higher is not alarming unless it exceeds the nameplate current rating of the motor. However, higher current means a slightly higher insulation temperature, which in turn means a slightly shorter life. If this goes on over a period of years, it becomes additive and motor life may be shortened substantially.

Low voltage can cause some other negative effects. Starting, pull-up, and pull-out torque of induction motors will be reduced. A reduction of nameplate voltage from 100 to 90% reduces these torque characteristics by a factor of 0.9 x 0.9 = 81% of their full-voltage values. As a result, hard-to-start loads will become difficult or impossible to start.

High voltage. The major effects of higher-than-rated voltage applied to an induction motor are higher starting current, increased starting torque, and lower power factor. There is an erroneous belief that since low voltage increases the current on a motor, then high voltage would tend to reduce current draw and the attendant overheating. This is not the case. Higher-than-

Brakes are fitted on to the back of motors when equipment must stop as soon at the motor starter is deenergized. A typical application is in conveyor equipment where coasting is unacceptable.

Code Letter	KVA/HP Range	Approximate Midrange Value
A	0.00 - 3.14	1.6
B	3.15 - 3.54	3.3
C	3.55 - 3.99	3.8
D	4.00 - 4.49	4.3
E	4.50 - 4.99	4.7
F	5.00 - 5.59	5.3
G	5.60 - 6.29	5.9
H	6.30 - 7.09	6.7
J	7.10 - 7.99	7.5
K	8.00 - 8.99	8.5
L	9.00 - 9.99	9.5
M	10.00 - 11.19	10.6
N	11.20 - 12.49	11.8
P	12.50 - 13.99	13.2
R	14.00 - 15.99	15.0
S	16.00 - 17.99	17.0
T	18.00 - 19.99	19.0
U	20.00 - 22.39	21.2
V	22.4 - and up	—

Table 1.1. Code letter designations and their values.

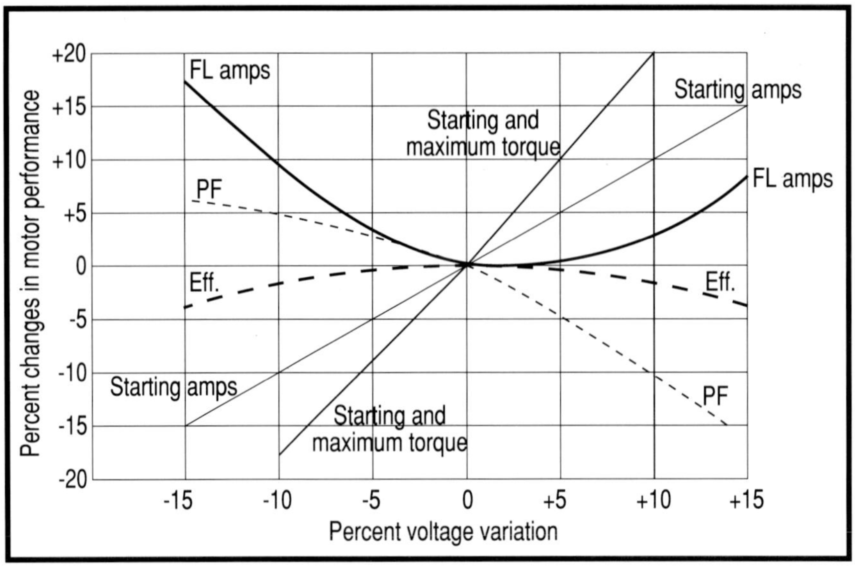

Fig. 1.6. Voltage variations change characteristics of a typical T-frame motor.

rated voltage tends to push the magnetic parts of the motor into saturation. This causes the motor to draw excessive current in an effort to magnetize the iron beyond the point to which it can easily be magnetized. Motors will tolerate a change in voltage above the design voltage, but extremes above the design voltage will cause the amperage to go up with a corresponding increase in heating and a shortening of motor life.

Fig. 1.6 illustrates the general effects of high and low voltage on the performance of T-frame motors. There is a great deal of variation from one motor design to the next, so the curves are only a typical example.

INDUCTION MOTOR STARTING CHARACTERISTICS

Locked-rotor current is defined as the steady-state current taken from the line with the rotor locked and with rated voltage and frequency applied to the motor. Locked-rotor currents for different type motors will vary from 2 1/2 to 10 times their full-load current; but there are motors with even higher inrush currents.

To define inrush characteristics and present them in a simplified form, code letters are used. Code letters group motors according to the range of inrush values and express the inrush in terms of kilovolt-amperes (kVA). By using the kVA basis, a single letter can be used to define the low- and high-voltage inrush values on commonly used dual-voltage motors. These code letter designations are shown in **Table 1.1** [which corresponds with Table 430-7(b) of the NEC].

To determine starting inrush amperes, the code letter value (usually the midrange value is adequate), horsepower, and rated operating voltage are inserted in an appropriate equation. The code letter is obtained from the motor nameplate. Inrush current of 3-phase motors is obtained as follows:

$$I_{inrush} = \frac{(\text{code letter value} \times \text{hp} \times 577)}{\text{voltage}}$$

Because values in the table are given in kVA/hp, the constant "1000" is needed to convert to volt-amperes. The constant "577" is obtained by dividing 1000 by 1.73.

The equation can be further simplified by dividing the constant by the rated voltage. The following equations for 3-phase motors give approximate results.

For 200V motors:

$$I_{inrush} = \text{code letter value} \times \text{hp} \times 2.9$$

For 230V motors:

$$I_{inrush} = \text{code letter value} \times \text{hp} \times 2.5$$

For 460V motors:

$$I_{inrush} = \text{code letter value} \times \text{hp} \times 1.25$$

Torque and rotor resistance are closely related — the higher the rotor resistance, the higher the starting torque. This relationship holds true only up to a limit, beyond which a further increase in resistance causes the torque to decrease. Torque, however, is also affected by the flux in the air gap and the disposition and shape of the rotor slots and bars. Modifications in the design of squirrel-cage motors permit a certain amount of control of the starting current and torque characteristics.

Motors have been categorized by NEMA Standards into four classifications.

NEMA Design A motors are types with normal torque and normal starting current. They have a locked-rotor current that can be anywhere from 6 to 10 times full-load current and have good running efficiency and power factor, high pullout torque, and low slip. The torque is about 150% at start. Pullout torque is over 200% of full-load torque

NEMA Design B motors are types with normal torque and low starting current. They have a starting current that is about 5 times the full-load current. Starting torque, slip and efficiency are nearly the same as those of a Design A motor. Power factor and pullout

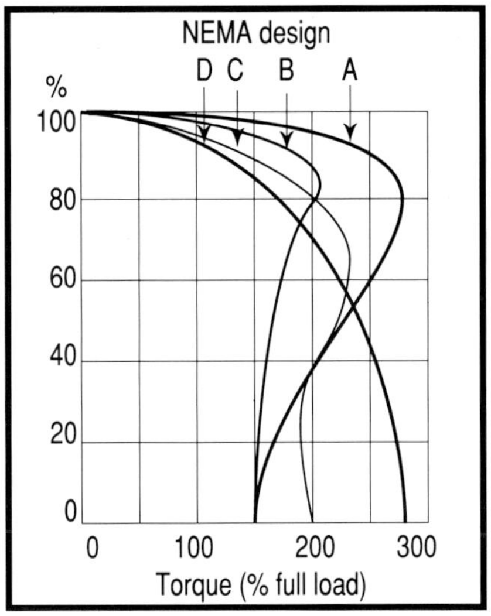

Fig. 1.7. Speed-torque curves for NEMA-design motors.

torque are somewhat less. This type is standard in 1- to 250-hp dripproof motors, and up to 100-hp TEFC motors.

NEMA Design C motors are high torque, low starting current, double squirrel-cage wound-rotor types. They have a higher starting torque (about 200%) and lower breakdown torque (about 180%) of full-load torque. Compared to both A and B designs, they have approximately the same full-load torque, and a lower current inrush than either. Typical applications are pulverizers, compressors, and conveyors.

NEMA Design D motors are high-slip types. They produce a very high starting torque (approximately 275% of full-load torque), and torque decreases continuously with speed. Starting current and efficiency are low and slip is high. These motors are used primarily where high starting torque is required but the running load is light or intermittent, such as for elevators and punch presses.

Fig. 1.7 shows the speed-torque curves for NEMA-design motors. When selecting a motor to drive a mechanical load, the designer can choose from among these categories the basic type of machine that will best meet the requirements of load and the nature of the electrical distribution system.

DC MOTORS

An ability to meet a wide range of torque and speed requirements makes DC motors suitable in a variety of applications. They are especially appropriate when smooth acceleration over a broad range, accurate speed change and/or speed matching, and close control of torque or tensioning are required.

Although AC induction motors controlled by adjustable-speed drives are readily available, the DC motor maintains its desirability in certain applications because of its special speed/torque characteristics. For example, AC motors driving heavy loads at about twice their rated torque will usually stall. DC motors, on the other hand, can deliver about three times their rated torque for short periods; and for very short periods (3 to 4 sec) they can deliver up to five times rated torque.

DC motors **(Fig. 1.8)** consist of two major components, the yoke or frame that contains the field windings, and the rotating component called the armature. An assembly consisting of a commutator that is part of the armature, and pairs of brushes that are part of the yoke are the other essential elements.

The yoke, also called the stator, is a cylindrical frame of high-permeability iron alloy to which are bolted the pole structures. The poles are arranged in an alternating north and south pattern, and the frame provides a return flux path as well as mechanical support for the poles, bearings, and brush holders. The stator poles can be either a permanent magnet or an electromagnet.

The armature consists of a shaft and laminated punchings of silicon steel slotted to accept the armature conduc-

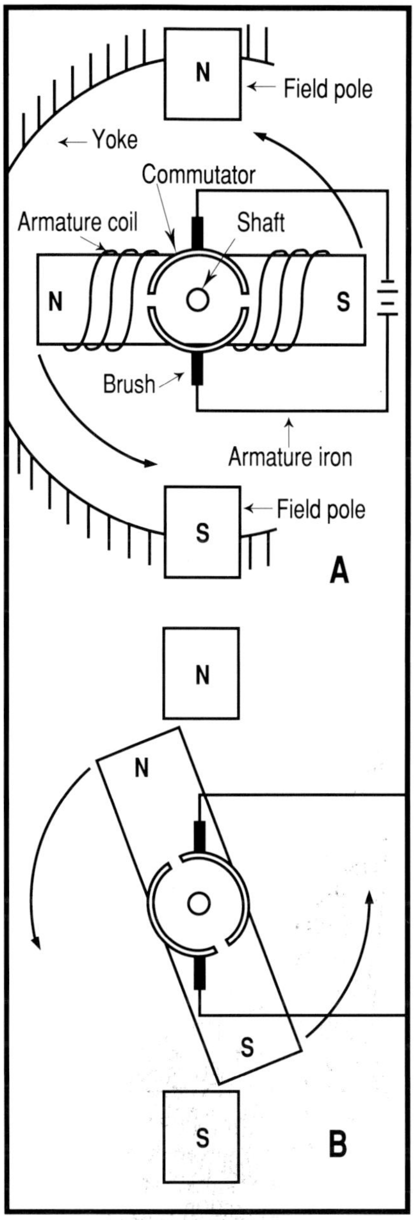

Fig. 1.8. The armature of a DC motor is rotated by alternate attraction (A) and repulsion (B)

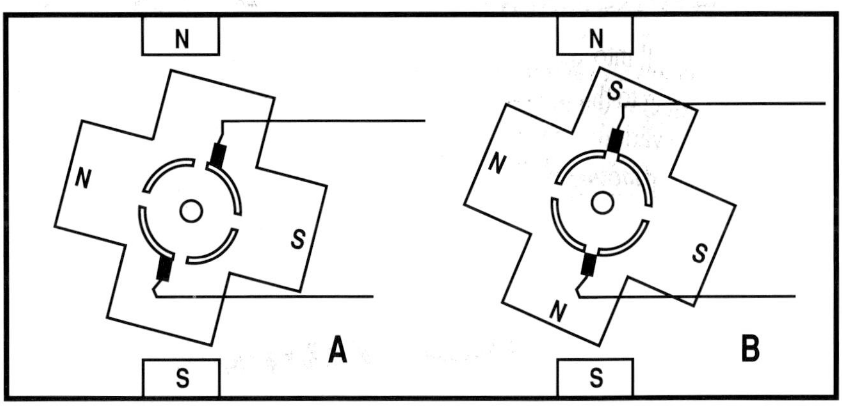

Fig. 1.9. Practical DC motors have more than two commutator segments and armature coils.

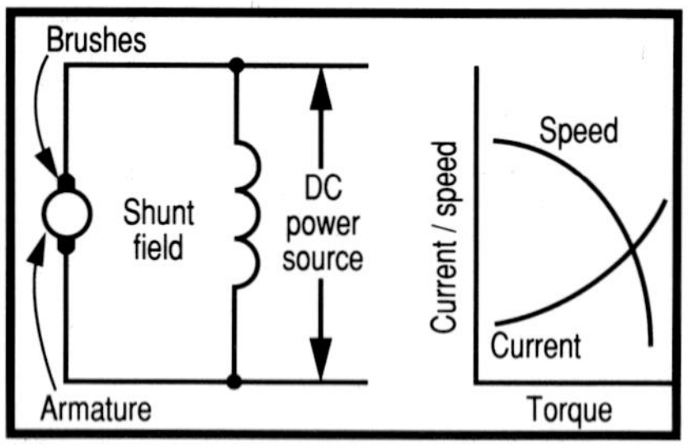

Fig. 1.10. Shunt-wound DC motor.

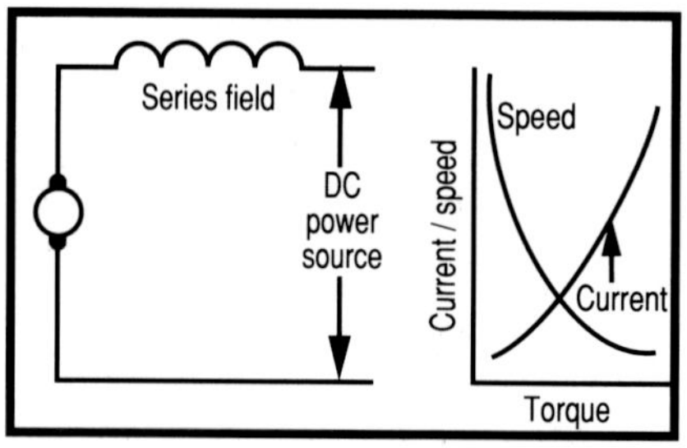

Fig. 1.11. Series-wound DC motor.

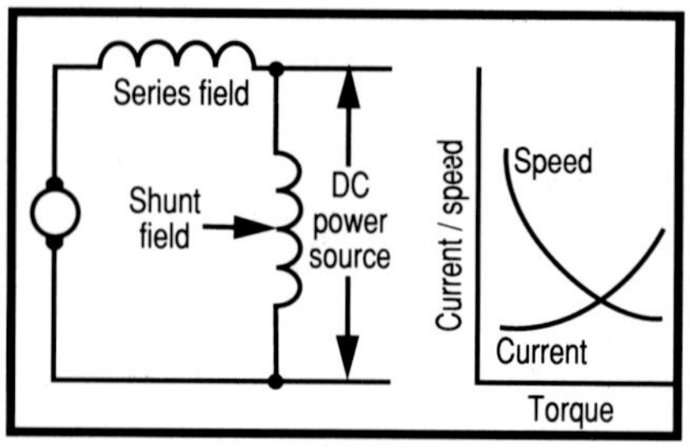

Fig. 1.12. Compound-wound DC motor.

tors or coils. The coil ends are connected to the segments of the commutator, which in turn connects the armature conductors to the power source through the brushes.

Armature voltage is provided from a DC source. Current flow from the source (battery or rectifier) in one direction through the brushes, commutator, and the coils creates an electromagnet with a south and a north pole. Since opposite poles attract, the north pole of the armature is attracted to the south pole of the stator, and the south to the north. Because the commutator is rotating with the armature, as the north and south poles reach virtual alignment, the brushes contact the other segment of the commutator, reversing current flow through the windings. The polarity of the armature poles is now reversed and its north pole is aligned with the north pole of the stator, and the south with the south. Like poles repel, forcing the armature to continue rotating towards the opposite pole.

Actual DC motors have more than two commutator segments and many more armature coils **(Fig. 1.9)**. With more commutator segments, it is not necessary to rely on inertia when the armature polarity is reversed. The segments change an instant before the poles of the fields are aligned. For example, in **Fig. 1.9(B)**, during the time that each brush contacts two segments of the commutator, four poles are created. However, all four poles provide torque in the same direction.

DC MOTOR CLASSIFICATIONS

DC motors fall into several classifications according to the types and connections of the various windings.

Shunt-wound motors **(Fig. 1.10)** are the most widely used types. The name originated from early operation of these machines with the armature and field circuits connected in parallel (shunt) to a constant-potential power supply. While the term "shunt" is still used, relatively few motors are now applied in this way. Shunt motors as now applied have their field circuits excited from a source of power separate from the armature power supply. The excitation voltage level is usually the same as the armature voltage, but special shunt-field voltage ratings of 15 to 600V are available as a modification. The shunt motor is characterized by its relatively small speed change under changing load; rarely will the drop-off exceed 5%.

The speed of the shunt-wound motor can be changed by varying the shunt-field current or armature voltage. Speed control by changing the armature resistance is unsatisfactory because the speed regulation is poor.

Whenever the load changes slowly, the flux changes as a result of the armature reaction and speed will remain constant. However, if the load changes more rapildy than the self induction of the field windings will allow the flux to change, then the speed will change rapidly. A speed range of approximately 4:1 with reasonable running stability is possible for loads up to full-load torque.

Care must be taken never to open the field of a shunt-wound motor that is running unloaded. The loss of field flux causes motor speed to increase to dangerously high levels.

Series-wound motors **(Fig. 1.11)** are connected so the field flux is produced by coils that are electrically in series with the armature. When the motor starts, the current, and consequently the magnetic field, are at maximum values, producing a large starting torque. As the motor speeds up and the current is reduced, the field flux is reduced. The torque will vary as the square of the armature current (neglecting saturation of the field poles), which reduces this relationship. The torque and speed are very sensitive to the load current (which is also the field current) because of the corresponding change in flux.

The speed of the series motor may be adjusted by shunting out the series winding, short-circuiting some field turns, or inserting resistance in series with the field and/or armature. How-

Resilient mounted split-phase motor (left) and general-purpose footless capacitor start motor (right) are among the various types or single-phase motors used in many different applications.

ever, speed adjustment is not easily accomplished. This type of motor has the disadvantage of tending to "run away" at light loads. The overspeed can reach a destructive value if the load is suddenly removed. For this reason, series-wound motors should be used only where the load is directly connected or geared to the shaft.

Compound motors **(Fig. 1.12)** have both shunt and series fields. By proportioning the relative amounts of series and shunt windings, the motor characteristics can the shifted to be more nearly shunt or more nearly series in nature.

Each winding has turns and wire sizes similar to the shunt-wound and series-wound motor field windings. The proportion of the total flux supplied by the series winding determines the amount of "compounding," which can be varied to suit the speed characteristics desired. A strong series field will give speed characteristics approaching those of a series motor. A weak series field will give characteristics approaching those of a series motor. Motors with series fields producing 40 to 75% of the total flux are often used, with a value of 50% being the most common. Compound motors having series fields producing 10 to 25% of the total flux are also used for some industrial applications.

Generally, the speed characteristics of compound motors lie between those of shunt-wound and series-wound motors. They can be used when speed variation with load variation is permitted.

The starting torque of the compound-wound motor is high, although not so high as that of a series-wound motor. The torque will increase rapidly with load because the series field will increase the flux. Speed will decrease rapidly for the same reason. However, the motor will not run away at light loads because of the shunt-field flux.

Speed of the compound-wound motor can be adjusted with a shunt-field rheostat. Modern DC motor controls for these motors are somewhat complex; however, they provide excellent control.

TEFC multispeed motors are available for use where the application calls for two fixed speeds that will be used at various times during an operating cycle, such as for fans and agitators.

ENCLOSURES

NEMA Standard MG1-1.25 classifies, in accordance with the degree of protection offered, open motors with ventilating openings that permit passage of external air over and around the machine windings.

- A- Dripproof
- B- Splashproof
- C- Semiguarded
- D- Guarded
- E- Dripproof, fully guarded
- F- Open, externally ventilated
- G- Open, pipe ventilated
- H- Weather protected (WP-I, -II)

NEMA Standard MG1-1.26 classifies, in accordance with the techniques utilized, totally-enclosed motors that prevent free exchange of air between the inside and outside of the enclosure.

- A- Totally enclosed, nonventilated (TENV)
- B- Totally enclosed, fan cooled (TEFC)
- C- Explosionproof
- D- Dust-ignition-proof
- E- Waterproof
- F- Totally enclosed, pipe ventilated
- G- Totally enclosed, water cooled
- H- Totally enclosed, water-air cooled
- I- Totally enclosed, air-to-air cooled
- J- Totally enclosed, fan cooled, guarded

FACTORS AFFECTING MOTOR PERFORMANCE

Single phasing. When a 3-phase AC motor operates under single-phase conditions, the voltages and currents are no longer three simple balanced sine waves separated by 120 electrical degrees. Instead, the voltages and currents are unbalanced, the phase angles vary considerably from 120 apart, and there is substantial harmonic content.

It has been shown mathematically that any set of unbalanced voltages or currents can be resolved into a number of balanced components, and that the total effect of the unbalanced voltages or currents is the same as the sum of the effects of the individual balanced components. This method is known as resolving the complex waveforms into "symmetrical components."

The three symmetrical components used in this analysis are: the positive-sequence component, the negative-sequence component, and the zero-sequence component. The positive-sequence current is that component of the total current that causes the motor to rotate in the proper direction; the negative-sequence component attempts to rotate the motor in the reverse direction; and the zero-sequence component is that current that flows in the ground circuit. In motor single-phasing, the

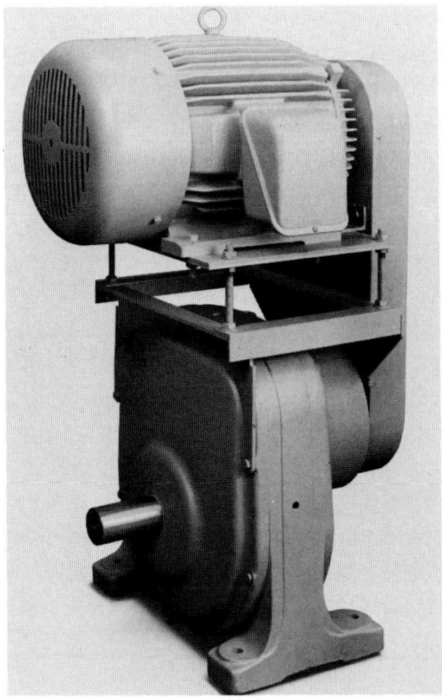

Mechanical speed variators are used in certain applications where the speed of the motor is only changed at long intervals, and a variable-speed or DC drive are not cost effective or practical.

negative-sequence component is usually smaller than the positive-sequence component, and the zero-sequence component is not a factor.

A motor can be depicted by an equivalent circuit (**Fig. 1.13**), with the rotor represented as a resistance. The positive-sequence current component creates a rotating magnetic flux that produces torque in the normal direction. The smaller negative-sequence current component has the same effect as if it creates a flux rotating in the reverse direction about two times as fast as the positive-sequence flux, and it produces torque in the opposite or reverse direction. The effective resistance of the rotor is anywhere from three to eight times as great for the negative-sequence current as for the positive-sequence current. Thus, each ampere of negative-sequence current creates three to eight times as much heat in the rotor as each ampere of positive-sequence current.

Since the resistance of the overload element is the same for both positive- and negative-sequence currents, the greater heating effect of the negative-sequence current on the motor is not seen by the OL. Thus, the motor can be damaged by heat before the overload relay responds and takes the motor off the line.

Voltage unbalance. Single-phasing can be considered a special case of voltage unbalance. Actually, any voltage unbalance will cause increased heating in the motor and is undesirable. This heating results from negative-sequence currents in the same manner as in single-phasing, but to a lesser degree. Unless the voltage unbalance is severe, it will not cause immediate motor failure, but will raise the motor operating temperature beyond design limits at full load. The percentage increase in temperature rise of the highest-current winding will be about two times the square of the percentage voltage unbalance. For example, a 3.5% voltage unbalance will cause about a 25% increase in the winding temperature rise. Thus, unbalanced voltages will ultimately cause premature motor failure. The greater the unbalance, the sooner the insulation will fail.

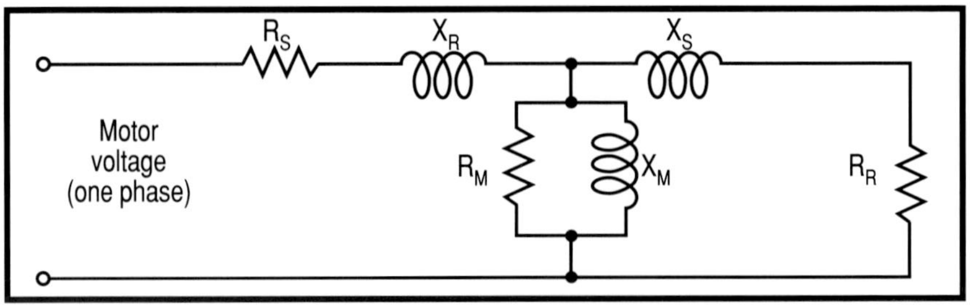

Fig. 1.13. Equivalent circuit for a motor. R_S is the stator resistance; X_S is the stator leakage reactance; R_M is the magnetizing loss equivalent resistance; X_M is the magnetizing reactance; X_R is the equivalent rotor leakage reactance; and R_R is the equivalent rotor resistance.

Harmonics. Harmonic voltages and currents are those at any integral multiple of the fundamental frequency. For example, in a 60-Hz system, the 2nd harmonic is 120 Hz, the 5th harmonic is 300 Hz, the 11th harmonic is 660 Hz, and so on. It has been shown mathematically that any periodic waveform can be resolved into a fundamental frequency and various harmonics. Therefore, any distorted 60-Hz current or voltage is considered to be the sum of the basic 60-Hz current or voltage and some combination of its harmonics (**Fig. 1.14**).

In a motor, the heating effect of a harmonic current is much greater than that of the same number of amperes of 60-Hz fundamental current. The higher frequency causes greater eddy current and hysteresis losses and heating in the steel laminations. The higher-frequency magnetic field drives the current to the outside of the conductors, increasing the effective resistance of the windings and the I^2R heating.

Some odd-order harmonics, especially the 5th, 11th, and 17th, cause a negative-sequence flux, resulting in greater rotor heating in the same manner as from unbalanced voltages. The result is that at a given load, a motor supplied with power containing harmonic distortion will run hotter than a motor supplied with pure 60-Hz power. However, a sine wave with harmonic content will have the same heating effect on the resistance heater of the OL, which will, therefore, provide inadequate protection.

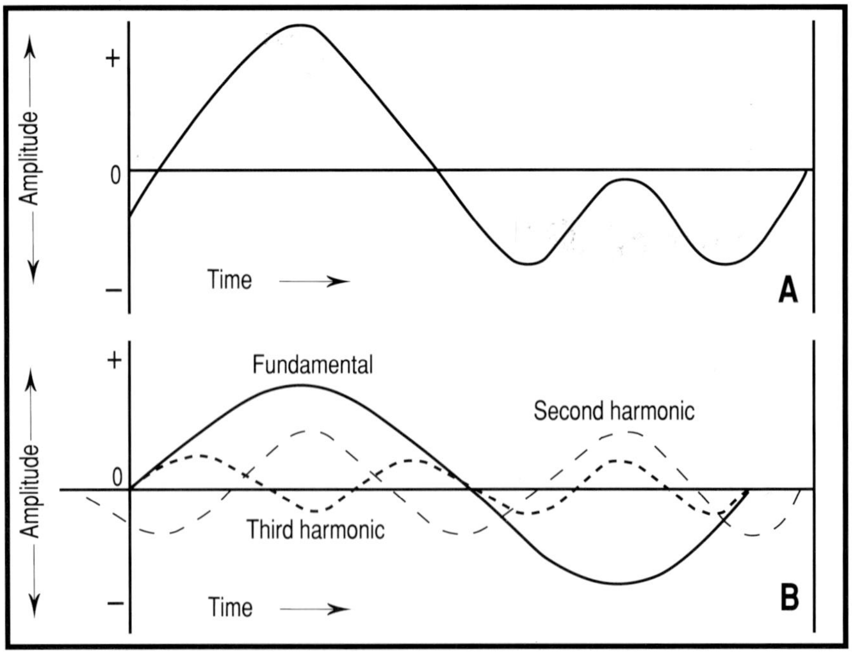

Fig. 1.14. A distorted sine wave (A) is generated by summing the values of the fundamental, 2nd harmonic, and 3rd harmonic (B) at each point along the time axis.

MOTOR CONTROLLERS AND DRIVES

2

STARTING AND STOPPING or otherwise regulating motors is the function of motor controllers and drives. There are as many different types of controllers (starters) and drives as there are types of motors. However, the most common types are those used to control squirrel-cage induction motors and DC motors.

ELECTROMECHANICAL STARTERS

Squirrel-cage induction motors are the mainstays of the electrical industry. The most common type used for starting these motors is the electromechanical across-the-line magnetic starter. As shown in **Fig. 2.1**, the unit consists of a contactor and a separate control circuit for closing and opening the contacts in the motor power circuit. In addition, these types of starters contain a thermal overload relay for protection of the motor against excessive loads, locked rotor, and other conditions that would cause excessive heat to be developed within the motor.

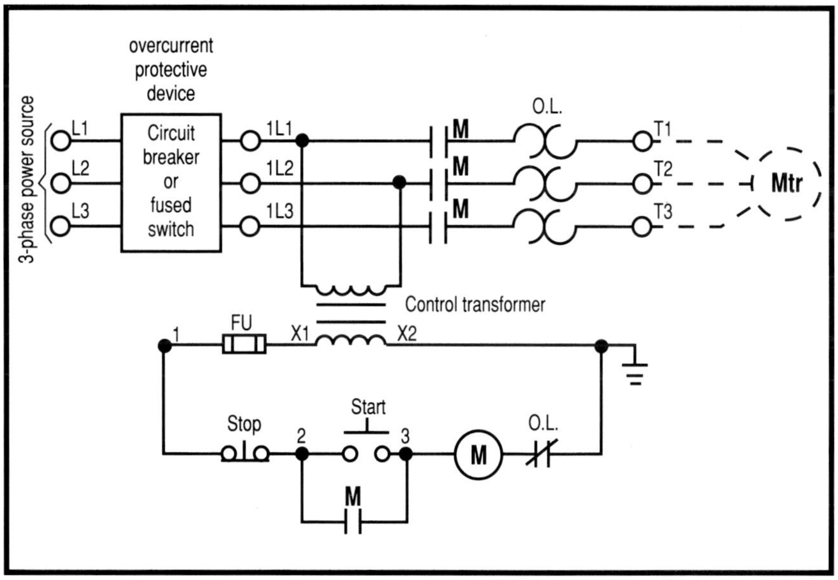

Fig. 2.1. Typical elementary ladder diagram of an across-the-line (full-voltage) combination starter.

Combination starters, which are most often employed, also contain a circuit breaker or fused switch that protects against short-circuit faults. The same protective device most often also acts as the motor branch circuit disconnect device.

Electromechanical starters are available in a large variety of types to meet the specific applications. Besides the most common, the across-the-line type, full-voltage starters also include reversing types **(Fig. 2.2)** and two-speed types. The two-speed starter is available for both separate winding motors **(Fig. 2.3)** and consequent pole motors. The consequent types are further broken down into those for constant torque or variable torque **(Fig. 2.4)** and for constant horsepower **(Fig. 2.5)**.

Selection of electromechanical starters requires an understanding of both the voltage at which it is to be applied, and the nameplate rating of the motor it is to control.

NEMA starters. Motor starters that

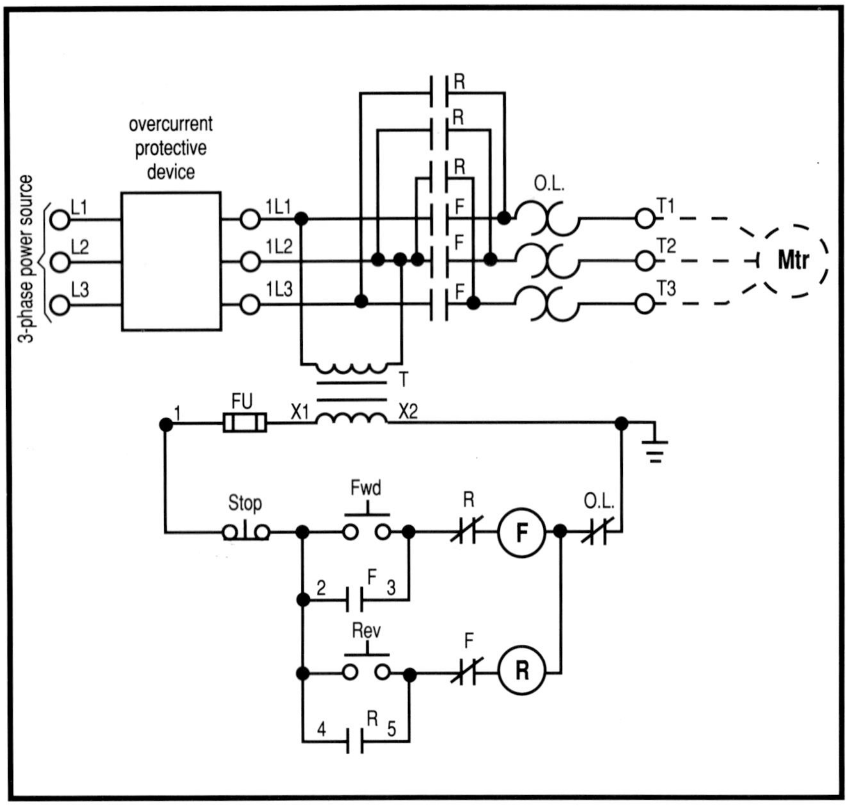

Fig. 2.2. Typical elementary ladder diagram of a full-voltage reversing combination starter.

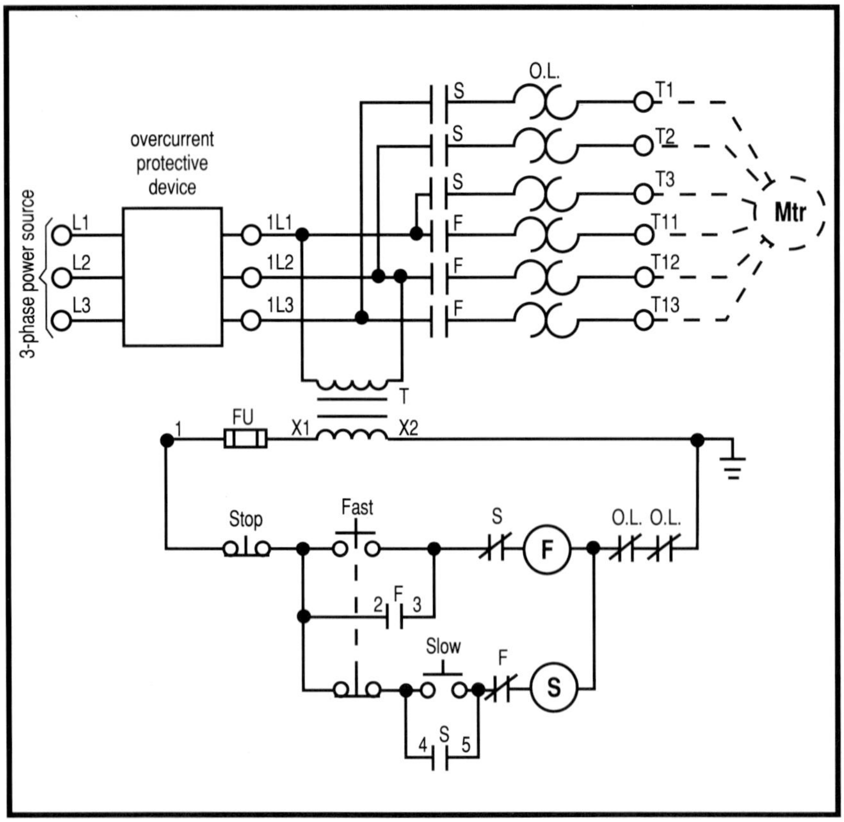

Fig. 2.3. Typical elementary ladder diagram of a multi-speed motor combination starter for separate-winding motors.

are manufactured according to the National Association of Electrical Manufacturers (NEMA) Standards are usually tabulated in such a fashion that if the voltage and hp are known, the appropriate "NEMA size" starter can be selected. Selection of the control voltage desired also determines whether the starter will require a control transformer to reduce the voltage within the control circuit. An option is to use an outside source of reduced voltage. Selection of overload protection generally is accomplished by selecting a properly-sized OL heater from a table supplied by the starter's manufacturer.

IEC starters. Starters that are manufactured to the International Electrotechnical Commission (IEC) Standards require a somewhat different method of selection. They are more compact and more load specific. Thus, there are more different sizes of starters in a series to cover the range of motor horsepower. The motor continuous current rating, and the voltage at which it is applied are the prime factors in the selection. Overload relays are provided specifically for each size of starter, and have an external adjustment to tailor the protection to the specific motor being used.

Both NEMA and IEC starters can be purchased as open types, or with enclosures to suit the environment. NEMA has assigned a designation to each according the type of atmosphere it is to be applied in. IEC starters have enclosures with an equivalent identification system.

- NEMA 1 enclosures are general purpose.
- NEMA 4 enclosures are watertight and dusttight.
- NEMA 4X enclosures are the same, but are also corrosion resistant.
- NEMA 7 & 9 enclosures are for hazardous locations.
- NEMA 12 enclosures are dusttight and driptight.

Starters are available in either individual enclosures, or they can be grouped within sheetmetal enclosures such as motor control centers (MCCs), power distribution panelboards, or busbar panels. The advantage of grouping is that much of the interwiring required between units can be done in

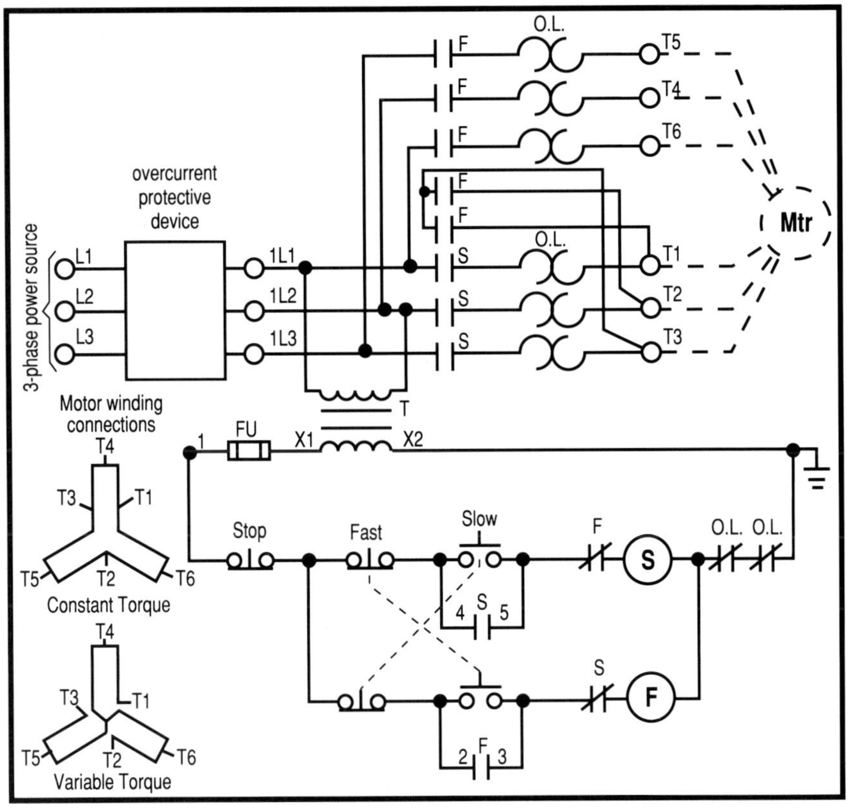

Fig. 2.4. Typical elementary ladder diagram of a multi-speed motor combination starter for consequent-pole motors (constant or variable torque).

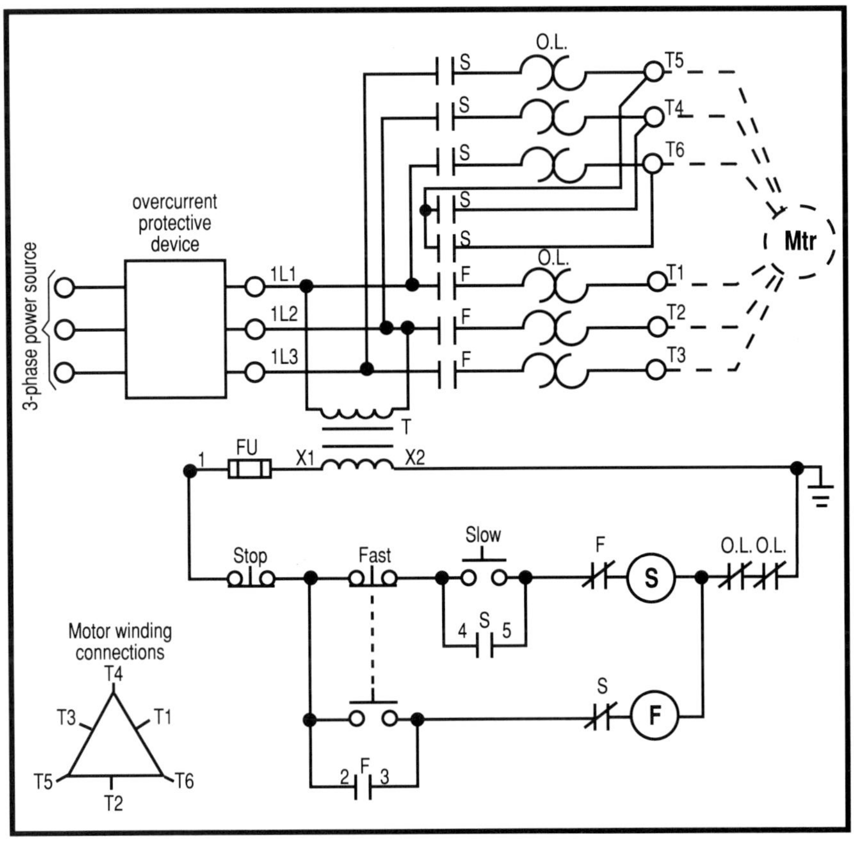

Fig. 2.5. Typical elementary ladder diagram of a multi-speed motor combination starter for consequent-pole motors (constant horsepower).

the factory. Also, programmable controllers, solid-state starters, variable speed drives, panelboards, and many other items can be located within the same enclosure.

ELECTROMECHANICAL REDUCED-VOLTAGE STARTING

When a motor is started at full voltage (across-the-line), the current in the motor will rise to about six to seven times its rated full-load current. During the acceleration time, it will have a low power factor. Uncontrolled inrush current can produce voltage dips on the power distribution system of the facility that can be very disruptive. Reduced voltage starting controls inrush current and limits it to values that can be safely handled. Flicker and other signs of lowered voltage levels are eliminated.

Torque during the starting period may be as much as two to four times normal running torque. It can cause undesirable mechanical shock to the motor and the driven machinery. Since motor torque is proportional to the square of the voltage supplied, starting at a reduced voltage will lower starting torque (and starting shock). For example, a 50% reduction of input voltage results in approximately a 75% reduction in motor torque. Lower starting torque means smoother acceleration of loads, resulting in less wear and tear on couplings, belts, gears, and other equipment being powered by the motor.

Various methods of reduced-voltage starting have been developed for use with squirrel-cage AC induction motors. Standard motors can be used with autotransformer, primary reactor, primary resistor, and solid-state starters. In addition, dual-voltage motors can usually be utilized with part-winding starters. Wye-delta motors and part-winding motors for the higher voltages (460 or 575V) are made to order. This raises the cost of the motor over a standard, readily available motor suitable for use on other types of reduced-voltage starting.

Table 2.1 lists reduced-voltage starter types and the results that can be

Starting Method	% of Full-Voltage Value		
	Voltage at Motor	Line Current	Motor Output Torque
Full voltage	100	100	100
Autotransformer 80% tap	80	64	64
65% tap	65	42	42
50% tap	50	25	25
Primary reactor 80% tap	80	80	64
65% tap	65	65	42
50% tap	50	50	25
Primary Resistor (typical)	80	80	64
Part winding	100	70	50
Wye start – Delta run	100	33	33

Table 2.1. Reduced-voltage starting methods for AC induction motors. The line current shown for autotransformer starting does not include magnetizing current, which can run as high as 25% of motor FLC.

expected in terms of motor voltage, line current, and the output torque of the motor.

In applying reduced-voltage starters, it is necessary to know the type of load that is being driven. For example, a centrifugal pump normally is very easy to start and can be operated with wye-delta starting or part-winding starting. These starting methods produce 33 and 50% of rated motor-starting torque respectively. They could also be expected to start an unloaded compressor. However, they would have difficulty starting a loaded inclined conveyor or a positive displacement pump because of high starting torques required on these types of loads.

In all cases, reduced-voltage starters cost substantially more than full-voltage (across-the-line) starters. The best starting method has to be one that achieves the desired result in inrush-current reduction and yields adequate starting torque to reliably start the load.

SOLID-STATE STARTERS

In recent years, the solid-state reduced-voltage starter has become available. With this type of equipment, the voltage is gradually raised electronically from zero up to the point at which the motor starts to turn the load. It continues to rise from that point to the final across-the-line operating voltage.

These types of starters are the modern version of the traditional electromechanical reduced-voltage starters. The primary function of a solid-state starter, in most instances, is to reduce motor starting current to a level that results in the least disturbance to the power distribution system. It also significantly reduces shock to the motor and load.

Other special functions offered by modern solid-state starters include: reversing capability, precise overload and overvoltage protection, phase-loss protection, jog, energy-saving features, and many others. In most instances when compared to an electromechanical reduced-voltage starter, solid-state starters are more compact, more versatile, and more reliable.

Sizes range from fractional-horsepower units to controllers that handle motors rated up to 700A at 460V.

How they work. A simple analogy helps in understanding the solid-state starter. **Fig. 2.6A** shows a motor circuit with 3-phase power supplied to the motor through power potentiometers. Placing the entire resistance of the three pots onto each of the three lines feeding the motor introduces maximum resistance at the motor input terminals, resulting in minimum or zero voltage applied to the motor. The total bypass of the potentiometers (minimum resistance) applies full input voltage to the motor input terminals. Any setting of the potentiometer between the maximum and the minimum applies a certain voltage to the motor input terminals between zero and full voltage.

The circuit in **Fig. 2.6B** employs solid-state electronic devices called silicon-controlled rectifiers (SCRs). These devices, which essentially have the same function as the potentiometers, also control voltage to the motor except that they do so more efficiently. Each single SCR controls one half cycle of AC voltage; six together provide full-wave control of motor input voltage. Operation of the SCRs is accomplished electronically by applying voltage to the gate of each SCR.

The difference between the solid-state devices and the potentiometers is that the semiconductors are either "on" or "off." They cannot be adjusted between zero resistance and infinite resistance. However, they provide a continuously variable output voltage to the motor by control of the point in the AC wave at which they are turned on or off. A microprocessor "gates" (applies proper voltage to) each SCR at a precise time. The six SCR method allows full control of the entire AC cycle. The SCRs are switched on and off by the microprocessor to simulate a sine wave at a selected (or programmed) adjustable voltage level. The ultimate

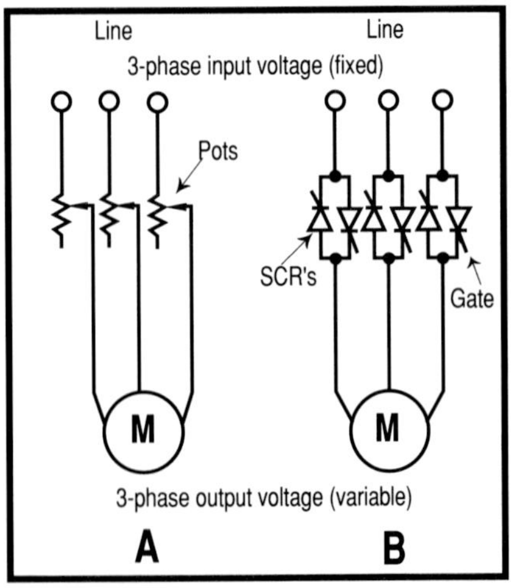

Fig. 2.6. Resistive control of motor voltage is shown in (A). As the potentiometers are adjusted from full-on to full-off, voltage to the motor varies proportianally. An equivalent method (B) uses SCRs to adjust the amount of time that voltage is applied to the motor.

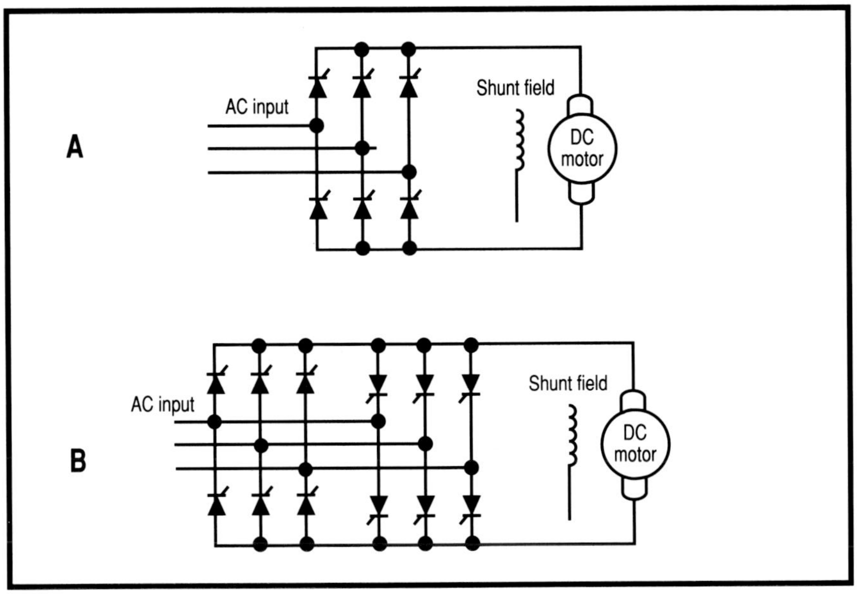

Fig. 2.7. Simplified DC controlled-rectifier drive unit (A), and a dual controlled-rectifier (regenerative) drive (B).

result is a continuously variable voltage to the motor. Adjustments for torque, time, and other parameters are made by a microprocessor to obtain switching of the SCRs to achieve the desired motor performance.

Components. A typical solid-state starter contains components for protection and control. They include: branch circuit protection/disconnecting means, normally a 3-pole circuit breaker; metal-oxide varistors to protect the SCRs and solid-state control circuits from overvoltages; a microprocessor module to provide a multitude of control functions; and overload protection similar to that used with electromechanical starters.

Settings. Controlled starting and reduced current can be achieved by providing sufficient torque to start and accelerate the load. Solid-state starters can be "fine-tuned" at the installation site to get just enough torque to do this at minimum current draw.

Solid-state starters provide smooth, stepless starts through the entire acceleration time, which is programmed in accordance with preset adjustments. Most solid-state starters provide at least two main adjustments: torque and time. Using the torque adjustment, the unit is set so that only sufficient torque is developed to start the motor rotating. The time adjustment allows for setting the amount of time needed for the motor to reach full power from the initial torque setting. This affects the motor acceleration time, with some variation dependent on the motor load. These adjustments are usually made in advance (after check-out tests) by manual settings on the microprocessor module. The adjustments should be set so that the motor begins to move the load in the shortest time, yet provides for a smooth startup of the load. The startup time for most solid-state starters (the ramp time) can be varied from 0 to 45 sec or more if required.

ADJUSTABLE-SPEED DRIVES

Efficient and reliable, adjustable-speed drives are used for many loads, such as fans, compressors, and pumps having variable-torque requirements, and also for control of tension, torque, position, or other variables.

Due to the relatively low losses of solid-state conversion devices, these units offer power and cost savings that are significant when compared to mechanical-type adjustable devices such as fan dampers, throttling valves, adjustable belts and pulleys, gears, magnetic clutches, and hydraulic drives. An understanding of the various types of electrical/electronic speed-regulating devices is important to their proper application.

There are six basic types of adjustable-speed drives that are in wide use at this time.

- DC drive with DC motor
- Voltage-source inverter with induction motor
- Current-source inverter with induction motor
- Slip-energy recovery system with wound-rotor motor
- Load-commutated inverter with synchronous motor
- Cycloconverter drive for either a synchronous or an induction motor.

Evaluations between the various types of electronic drives are often attempted. However, a general comparison is not possible since efficiencies can vary between specific products in each type of drive. The motor design and specific operating points are the largest contributors to efficiency differences. Power circuit configuration must also be part of the calculations.

DC drive **(Fig. 2.7)** is the oldest and most used of the various types of adjustable-speed drives. Its power-conversion circuitry, which is the simplest of the electronic drives, provides continuous speed control over a given speed range. The speed control can be either manually set with an operator's potentiometer or automatically set from various input signals responding to the process being controlled.

Generally less expensive than AC drives in the low-horsepower range, the DC drive's major cost item is the motor. If any special motor modifications are required, the cost may increase considerably. The DC motor is larger and heavier, requires more maintenance (especially the commutator and brushes), and is less adaptable to hostile environments (such as dust, moisture, or combustible gases) than the equivalent AC motor. Generally, for single-motor noncritical speed-control service, the DC system is the proper selection. AC units are preferable for multiple-motor applications.

AC drives. Inverters are categorized by the way they synthesize an AC waveform. Step-type inverters are made as either current source or voltage source. Their operating concepts can be understood by looking at their voltage and current sine waves **(Fig. 2.8)**.

• The six-step square-wave current-source inverter **(Fig. 2.8A)** produces current steps whose amplitude is proportional to the selected frequency. Current is constant in each step with the net effect being a variation in the rough approximation of a sine wave. The line-to-line voltage waveform associated with this current shows six spikes per cycle. Phase shift results from the power-factor angle. The rectifier of the current-source inverter regulates current and frequency.

• The six-step square-wave voltage-source inverter **(Fig. 2.8B)** has the amplitude of the line-to-line voltage steps proportional to the selected frequency, producing a varying effective voltage. The inverter uses a phase-controlled rectifier to obtain the variable DC voltage level that determines the output voltage to the motor. Line current approximates a sine wave distorted by harmonics of the selected frequency.

• A variation of the voltage-source inverter, the voltage-source PWM inverter, **(Fig. 2.8C)** produces pulses with a constant-voltage amplitude. However, these square-wave pulses vary in number and width to provide a current amplitude that varies in proportion to the selected frequency. The current waveform has a sinusoidal shape on which high-amplitude harmonics at the switching frequency also appear.

HOW AC DRIVES WORK

To properly apply AC drives, it is important to understand how they work.

Current-source inverter shown in **Fig. 2.9A** controls the torque of a squirrel-cage induction motor by regulating current directly. The current-source inverter has a phase-controlled rectifier that provides a DC input to a six-step inverter. The filter for ripple reduction in the DC link is a choke or reactor that provides a high impedance to rapid changes in current. Control of the inverter serves to regulate current and frequency, rather than voltage and frequency as with the voltage-source inverter. With a six-step inverter, the motor stator field "jumps" six steps per 360° of rotation, so there are six impulses per revolution to the rotor of a 2-pole motor, 12 for a 4-pole motor, and 18 for a 6-pole motor.

For applications requiring large induction motors (typically over 700 hp), the drive equipment can incorporate a 12-pulse unit power-conversion section, comprised of two six-pulse six-step inverters.

As seen in **Fig. 2.9B**, the six-pulse sections are "connected" together, or paralleled, by an input and an output transformer. The top section is called the "master" and is connected to the incoming power supply through a delta/delta transformer. Its output also is connected to the motor through a delta/delta transformer. The "slave" section is an identical unit except that its input is connected to the incoming power supply by a delta/wye section of the transformer and its output section to a wye/delta section of the output transformer.

While a delta/delta transformer has no phase difference between the pri-

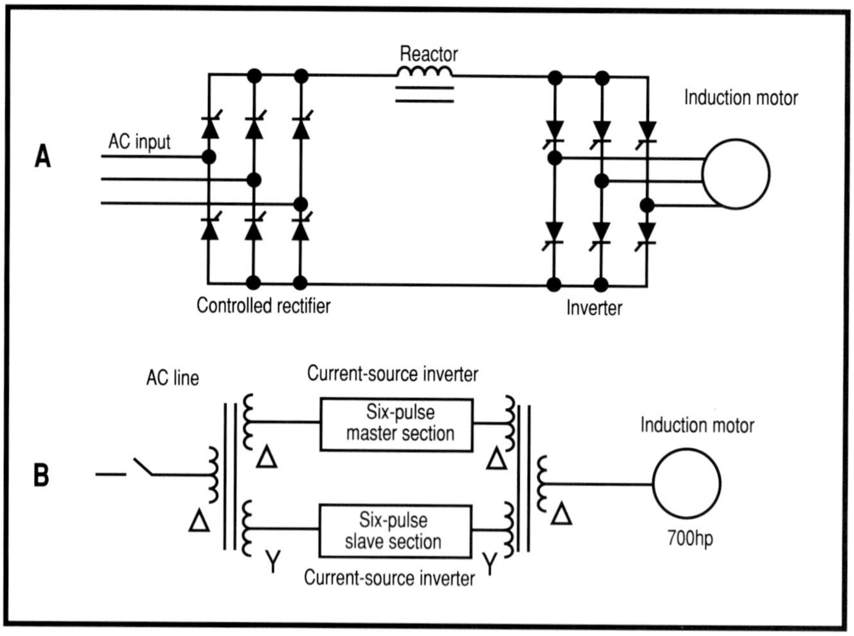

Fig. 2.9. Typical current-source inverter (A) and one with a 12-pulse power-conversion unit (B) required by larger motors.

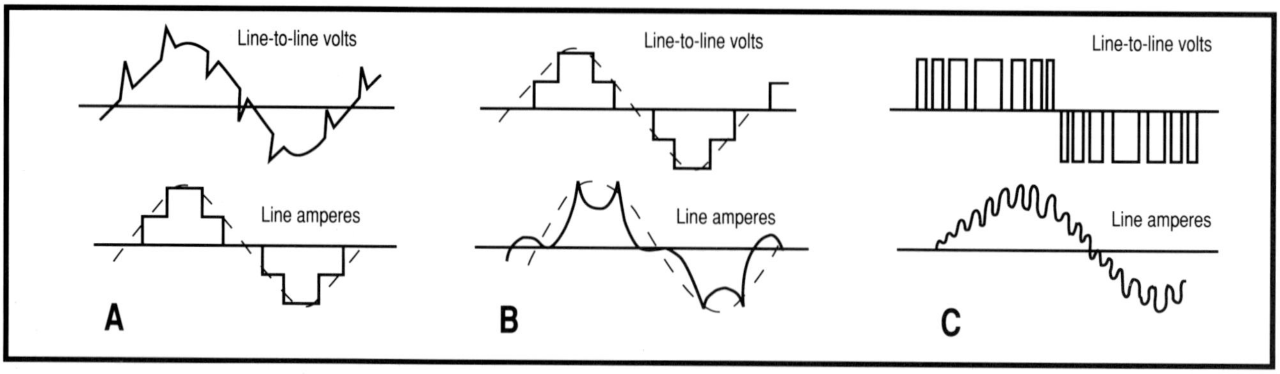

Fig. 2.8. Voltage and current waveforms. Current-source six-step inverter (A); voltage-source six-step inverter (B); and a pulse-width modulated inverter (C).

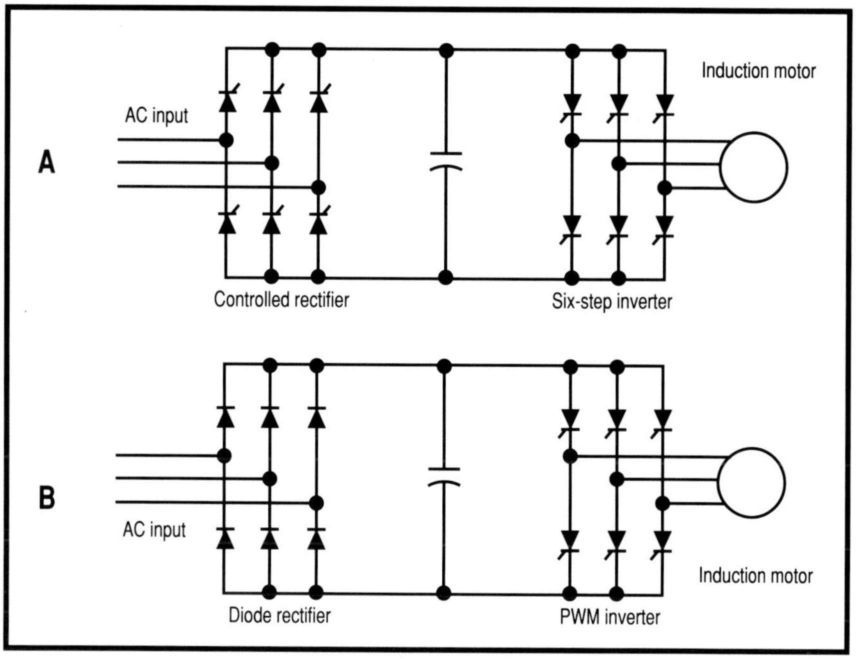

Fig. 2.10. Six-step voltage inverter (A) and a pulse-width modulated inverter (B).

mary and secondary, a delta/wye transformer has a 30° phase shift between its input and output. Thus, the master and slave currents are shifted 30° apart from each other. This resulting 12-pulse current waveform looks more sinusoidal than the six-step (pulse) current waveform, and the 12-pulse waveform has a lower harmonic content than the six pulse type (1% compared to 5%). On some applications, a dual-winding motor can be used, eliminating the need for an output transformer.

Voltage-source inverter regulates the output frequency to obtain the desired speed and ratio of voltage-to-frequency to achieve the desired motor torque. A constant voltage-to-frequency ratio approximates constant motor excitation (flux), which in turn allows constant output torque. At low frequencies, the voltage-to-frequency ratio is raised to obtain constant motor flux. The power circuit of a voltage-source inverter is made in different types:

- The six-step voltage-source type **(Fig. 2.10A)**, as does the current-source inverter, has a phase-controlled rectifier that provides a DC input to an inverter bridge. The difference is that a large filter capacitor is used in the DC link, rather than a series reactor. This provides a stiff voltage source to the voltage-fed inverter. A commutation circuit also is required.
- The pulse-width modulated (PWM) voltage-source type **(Fig. 2.10B)** has an input diode bridge that provides constant voltage DC to an inverter bridge. When very slow speeds are required, a PWM inverter is preferred to a six-step voltage source inverter because it produces more pulses and thus develops a smoother waveform. The PWM unit also offers better power factor than a six-step inverter.

Slip-energy recovery drive **(Fig. 2.11)** has a similar or even higher efficiency than other electronic drives. However, unlike them, the full motor power is not delivered through the

AC adjustable speed drives have wide application because they regulate the speed of the rugged squirrel-cage induction motors that are used extensively to drive pumps, blowers, and other loads.

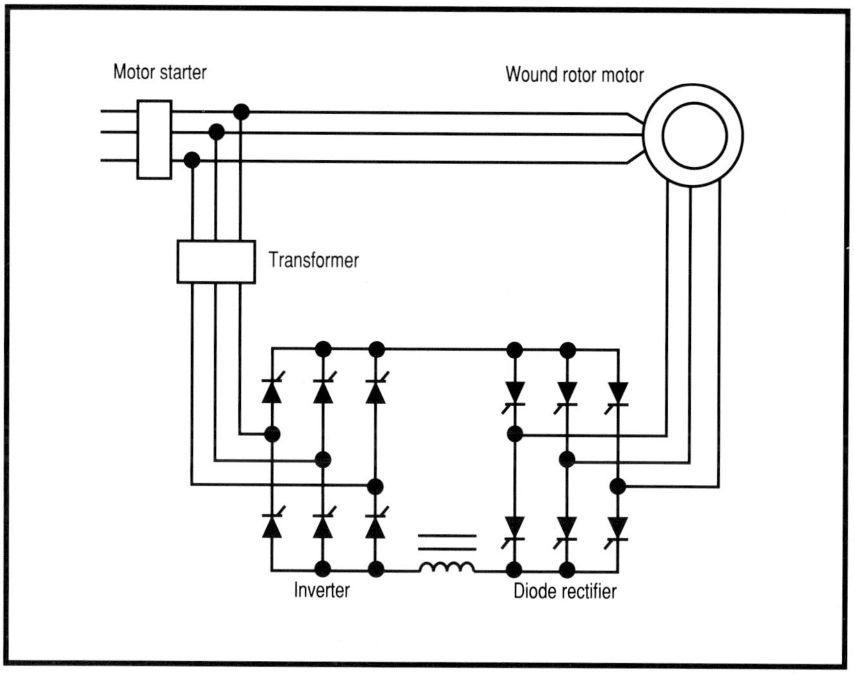

Fig. 2.11. Wound-rotor slip-energy recovery drive.

controller. Therefore, when a limited speed range is required, the controller can be sized significantly smaller than a speed-control device supporting the full-rated motor power. This recovery drive operates below synchronous speed and cannot operate above synchronous speed.

The major cost is the motor, which has speed control in the rotor circuit. In the past, wound-rotor, slip-ring induction motors used in adjustable-speed drives had variable resistors in the rotor circuit. The rotor energy was converted to heat in the resistors, and this heat was given up in the atmosphere, or cooling equipment was added to remove the heat. Since this was an inefficient design, the slip energy from the rotor circuit in today's device is rectified to DC and converted to a fixed voltage and fixed frequency to be fed back to the motor stator. The inverter for the slip-recovery unit is line commutating.

An AC power source supplies the energy to turn off the thyristors, similar to a DC drive operating in the regenerative mode. This eliminates the need for a separate commutating circuit.

Load-commutated inverter drive **(Fig. 2.12)** is a current-source AC drive serving a synchronous motor. The DC magnetizing current to the rotating-field windings is provided without resorting to slip rings by using a brushless-excitation method. The brushless exciter works like the alternator of an automobile. AC power is fed to the exciter stationary or primary windings and transformed to the rotating or secondary windings of the motor's rotor. The AC output of the secondary winding attached to the rotor shaft is rectified to a DC voltage by a full-wave silicon-diode rectifier, also built into the rotor, and delivered to the rotor-field windings. Since excitation for magnetizing the rotor is provided not from the inverter but separately, the drive is often referred to as a brushless DC motor drive.

Because the motor serves to commutate the inverter thyristors, the inverter does not require commutating capacitors. Thus, this type of drive is the most efficient one for adjustable-speed applications and is especially suited for high-horsepower motors. However, large salient-pole synchronous motors are more likely to have problems relating to electrical torque pulsations than induction motors.

Cycloconverter drive **(Fig. 2.13),** through a one-step conversion process, changes the frequency of the 60-Hz power input. The maximum output frequency applied to the motor, which can be an induction or synchronous type, is only a fraction of the supply frequency. The cycloconverter drive, which can equal the performance of a DC motor under current control, is normally used for very large horsepower, low-speed applications, providing high efficiency and fast response.

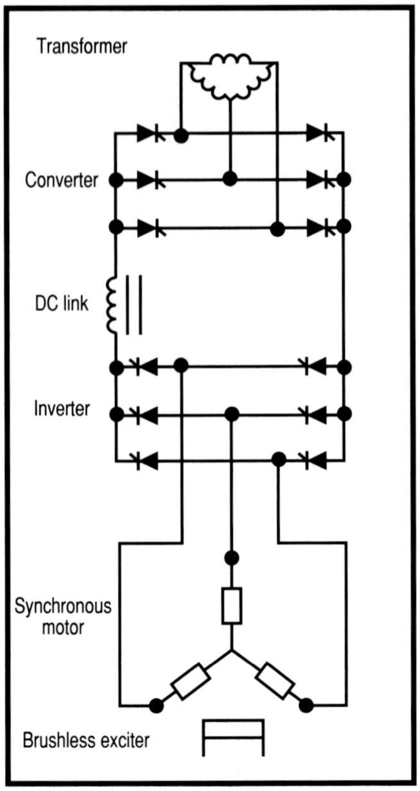

Fig. 2.12. Load-commutated inverter.

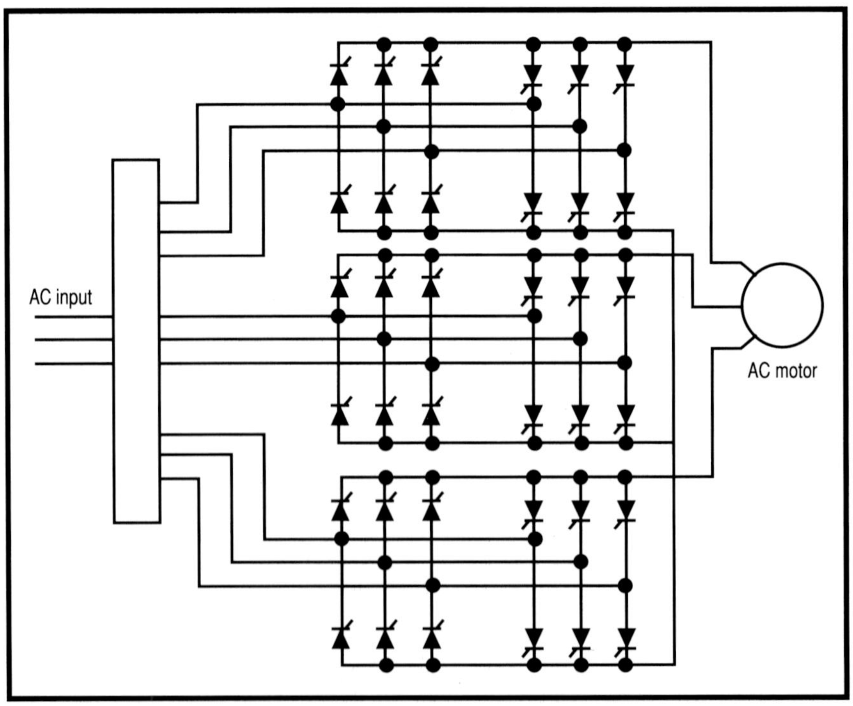

Fig. 2.13. Cycloconverter drive.

MOTOR OVERCURRENT PROTECTION

3

PROVIDING short-circuit and ground-fault protection for the motor, circuit conductors, and control apparatus is required by NE Code Article 430, Part D. The ground-fault protection called for in this part of the NEC refers to the protection against faults provided by a circuit breaker (CB) or fuse, not a zero-sequencing transformer and relay hookup.

Overcurrent protection is intended to prevent damage to conductors and insulation from excessive current that flows due to a line-to-line or line-to-ground short circuit. The argument about whether CBs or fuses provide the best protection has been going on for a long time. The truth is that there are some applications that only CBs can satisfy, others that only fuses can satisfy, and many in which either can perform satisfactorily. Sometimes a CB/fuse combination can provide the protection that neither can do alone.

MOLDED-CASE CIRCUIT BREAKERS

In motor controllers, circuit breakers often serve as the overcurrent protective device and as the disconnect means required by the code. In low-voltage circuits (less than 1000V), molded-case circuit breakers are the principal type of CB used. They can be divided into two classes: thermal-magnetic types, and magnetic-only types. Other types including fused circuit breakers, solid-state trip CBs, current-limiting CB, and motor circuit protectors are only variations of these two basic types.

Thermal-magnetic CB. The basic CB in use is thermal-magnetic in tripping action. As shown in **Fig. 3.1,** the current path within the breaker is through a bimetallic strip. The resistance of the bimetal develops heat, which

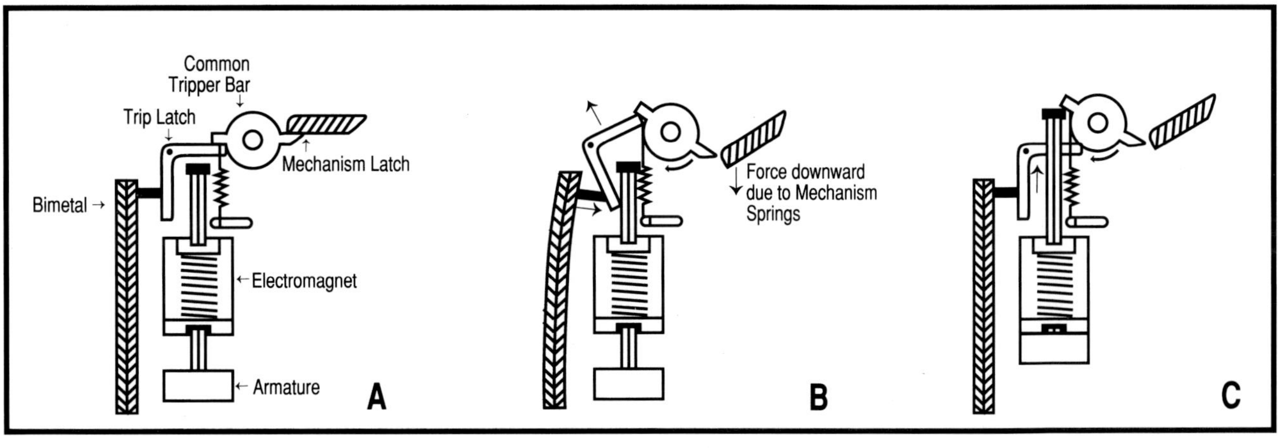

Fig. 3.1. Thermal-magnetic circuit breaker tripping action. (A) depicts the trip elements during normal current; in (B), the circuit breaker is tripped through the action of the thermal element; and in (C), the tripping action is caused by the magnetic-trip element.

causes the bimetal to bend until it moves far enough to unlatch the mechanism and allow the breaker to trip open. This action is the thermal tripping due to low-level overcurrent.

For high fault currents, thermal action is too slow. Very large, sudden increases in current builds up a magnetic field in a solenoid coil that is also in the current path within the breaker. This attracts a magnetic armature that strikes and unlatches the trip mechanism of the breaker, opening the circuit.

The thermal action provides inverse time response. That is, on small overcurrent values, it takes a long time for the bimetal to trip the breaker. The higher the overcurrent level becomes, the shorter the time needed to trip the CB. The magnetic-trip response is instantaneous; it either trips without time delay, or it does not trip at all. The result is shown in **Fig. 3.2** time-trip curve for a typical thermal-magnetic CB.

A thermal-magnetic CB is defined by its "frame size" and its "trip setting." Frame size is defined as a group of CBs with similar physical configurations. Frame size is expressed in amperes and corresponds to the largest ampere rating available in the group. Typical standard sizes are: 100A, 225A, 400A, 600A, 800A, 1200A, and so on. The standard ampere ratings of inverse-time CBs are set by the NE Code, with 15A being the lowest rating, and extending up to 6000A.

The same functions of the standard thermal-magnetic circuit breaker can be duplicated by molded-case circuit breakers that employ solid-state trips. The solid-state breaker uses sensors that respond to the current and provide a signal to the trip unit that is proportional to the magnitude of the current.

Magnetic-only CB. A thermal-magnetic CB provides poor motor circuit protection. Improved protection is afforded a by magnetic-only CB. It is essentially a thermal-magnetic CB that has had its thermal elements removed.

For specific application to motor circuits, the magnetic-only CB is often referred to as a motor-circuit protector (MCP). The MCP has the trip adjustments calibrated to ampere rating. Because it does not have the ability to sense low-level overcurrent, the MCP must be used in a combination starter that has an overload relay. The overload relay provides time delay for starting as well as overcurrent protection up to the locked-rotor current of the motor.

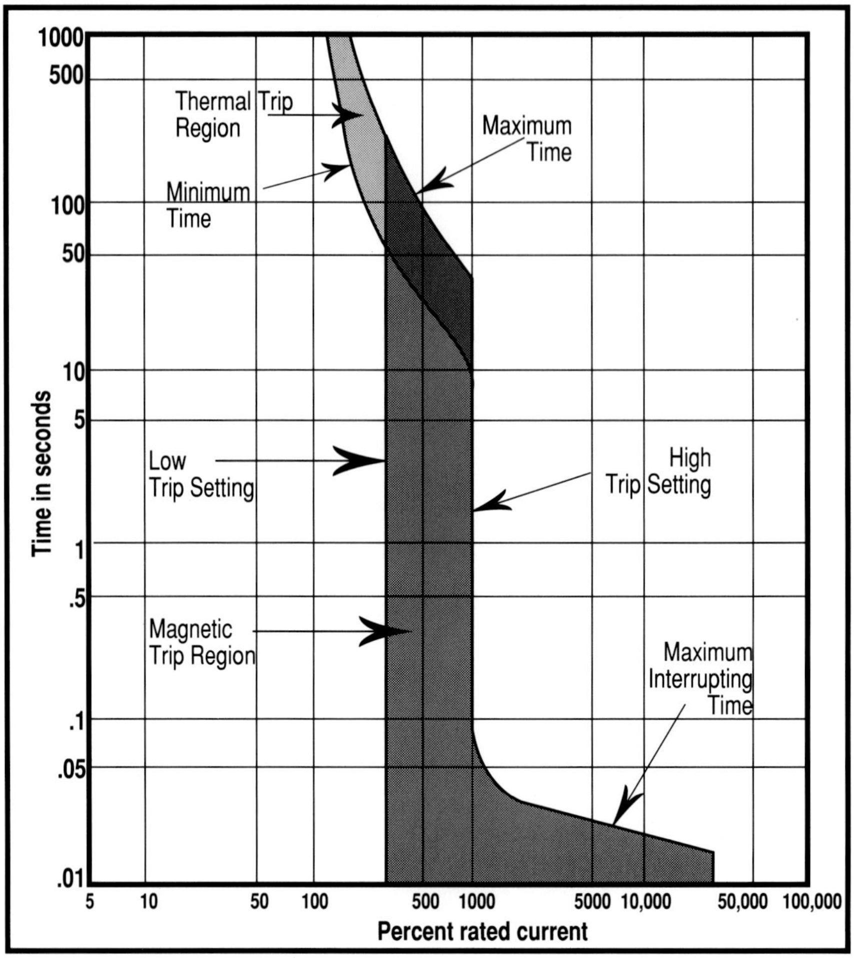

Fig. 3.2. Molded-case CB trip curve. The thermal-trip portion of the curve is fixed by the characteristics of the bimetal. Tripping can take place at any time between the minimum and maximum limits. The magnetic portion of the curve is a single vertical line that can be adjusted between the low and the high trip settings.

Magnetic-trip settings are adjustable. There is no specific ratings for MCPs that are mandated by the NE Code. Therefore, the appropriate size MCP can be set much closer to desired trip point that will protect the motor. This feature is especially useful in the lower-hp motors where the 15A minimum setting specified in the code for thermal-magnetic inverse-time circuit breakers is too high.

LOW-VOLTAGE FUSES

Modern fuses use zinc, copper, or pure silver elements with very carefully designed characteristics. They are highly accurate and can have high interrupting capacity or current-limiting ability. Properly selected and coordinated, they can do an excellent job of providing overcurrent protection.

They are catagorized by standards into various classes having certain specific dimensions and characteristics.

Class H fuses are the original cartridge fuses. They come in 250V and 600V dimensions, are ferrule types to 60A, blade-type above 60A to their maximum rating of 600A, and have a short-circuit rating of 10,000A.

Class K fuses are current limiting **(see Fig. 3.3)** but cannot be labeled as such because they have the same 250V and 600V dimensions as Class H fuses with which they can be physically interchanged. They are generally labeled "energy limiting" and have interrupting

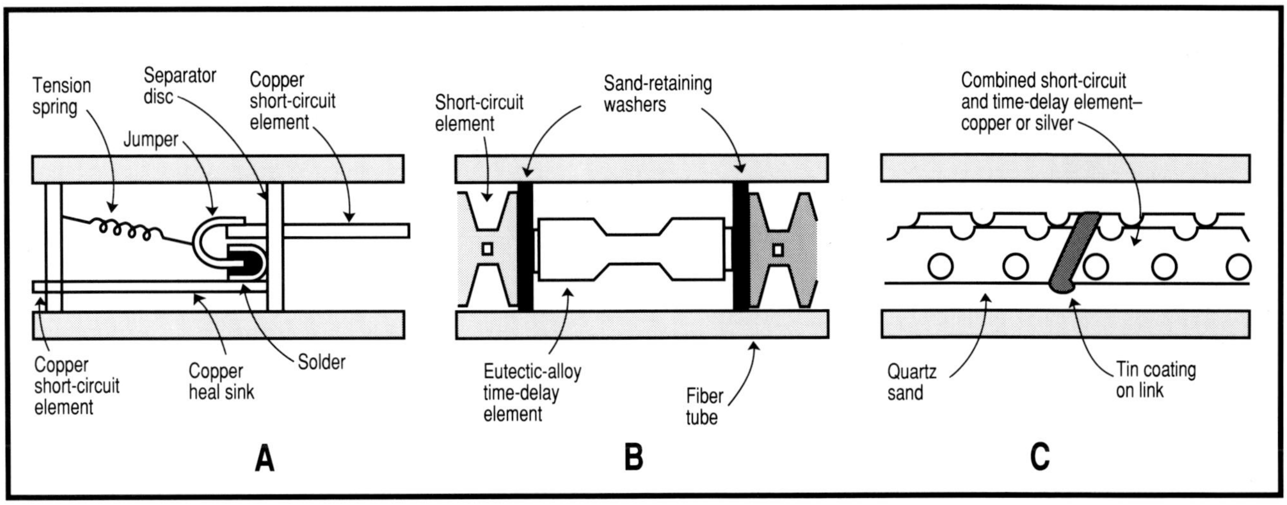

Fig. 3.3. Methods used to obtain time-delay action in fuses. In (A), a spring-loaded jumper, soldered in place completes the circuit within the fuse. As the element slowly heats with sustained overcurrent, the solder melts and the spring pulls the jumper out of position, opening the circuit. In (B), a sustained overcurrent heats a large element made of eutectic alloy until it reaches its melting point, at which time it drops away, opening the circuit. In (C), a single fuse element is used for both instantaneous short circuit and time-delay overcurrent action. On sustained overcurrent, a small tin overlay in the center of the link melts and increases link resistance, causing local heating until the link opens. Time-delay is obtained by heat transfer from the large surface of the folded link to the quartz-sand filler, overriding brief overcurrents.

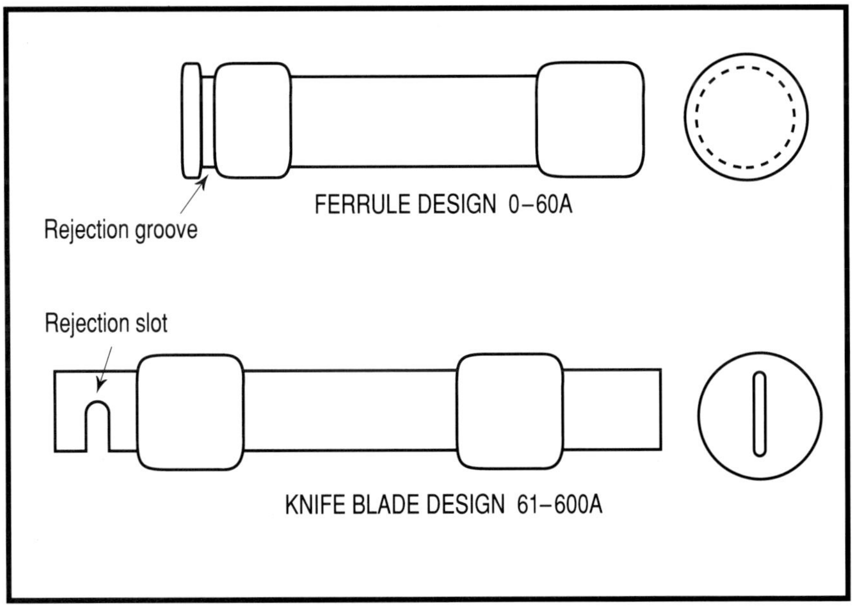

Fig. 3.4. Class R fuse noninterchangeable features.

ratings up to 200,000A. In Class K fuses, time delay is defined as the ability to carry 500% of its current rating for a minimum of 10 sec.

Class R fuses are current limiting and are labeled as such. They cannot be replaced by noncurrent-limiting fuses because of their special rejecting features as seen in **Fig. 3.4.** They are always rated 200,000A symmetrical interrupting capacity and are available up to a maximum continuous current rating of 600A.

Class J fuses also are true current-limiting fuses and are so labeled. They are rated 200,000A symmetrical interrupting capacity and limit peak current and I^2t let-throughs to specified maximum values. They are physically smaller than Class H, K, and R fuses. Class J fuses have bolt-holes on the blades and can either fit fuse clips or can be bolted in place. Noncurrent-limited fuses cannot be inserted in their place because of their dimensions. They are rated for 600V and are available in sizes up to 600A.

Class L fuses are listed for normal duty with continuous rating over 600A to 6000A. They are bolted in place and their bolt-hole configuration varies with the fuse rating. Class L fuses have a 200,000A interrupting rating.

Class T fuses are true current-limiting fuses, and fit in fuseholders that will fit no other fuse. They are made in 250V and 600V dimensions for current ratings up to 600A, and are physically smaller than Class J fuses. Class T fuses have no time-delay rating and have an interrupting rating of 200,000A.

Class G fuses are noninterchangeable, with ratings up to 60A at 300V maximum to ground. They have an interrupting capacity of 100,000A. Class G fuses are physically very small.

Fuses serve the same function in protecting against faults as circuit breakers. It is necessary for those applying fuses to coordinate their characteristics with that of other motor control devices that are employed in order to properly protect the circuit and components. Manufacturers provide three essential types of fuse data:

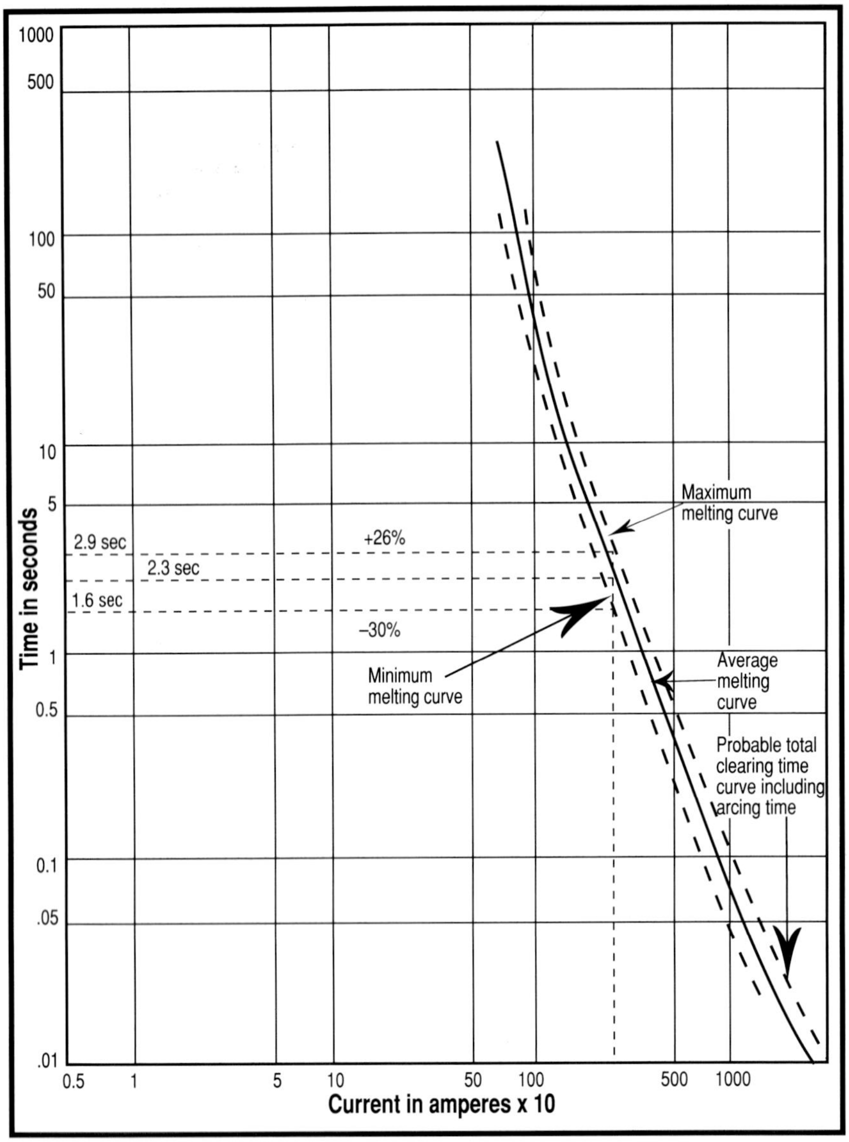

Fig. 3.5. Typical fuse curve. The melting band width (dotted) describes the tolerance on each side of the average melting curve. In this example, the average melting time for the fuse is 2.3 sec at that value of current. It can actually melt in a minimum of 1.6 sec or a maximum of 2.9 sec.

- time-current curves showing fuse melting times at various currents **(see Fig. 3.5);**
- current-limitation curves showing peak current let-through for various symmetrical available currents **(Fig. 3.6);**
- and curves or tables showing the I^2t let-through and damage levels for different fuses at specific available fault currents **(Fig. 3.7).**

In motor control circuits, a fused disconnect switch is often specified instead of a circuit breaker in a combination starter. Fuses are quick acting and provide assured circuit interruption in case of a fault that exceeds the rating of the protective device. The switch serves as the disconnect required by the code. On the negative side, the blowing of a fuse on only one leg of a three-phase motor causes single-phasing that can seriously damage the motor. Various strategies are used to overcome such possibilities.

RATINGS OR SETTINGS

The basic requirement for the rating or setting of the short-circuit and ground-fault protective device is that the device must be capable of carrying the starting or inrush current, and that it be rated or set at a value not exceeding that permitted by the NE Code.

In cases where calculated values do not correspond to the standard sizes of fuses or ratings of circuit breakers, the next higher standard rating or size may be used. However, where maximum ratings for branch circuit short-circuit protection are shown in the manufacturer's literature or are otherwise marked on the equipment packaging, they must not be exceeded, even though higher ratings or settings are permitted by the code.

Circuit components are required to be coordinated so as to permit circuit-protective devices to clear a fault without extensive damage to the electrical components of the circuit. While the wording "without extensive damage" is somewhat vague, the intent is to require coordination between the capabilities and limitations of the equipment to be protected and those of the selected protective device.

EXAMPLE. The motor circuit shown in **Fig. 3.8** has an available short-circuit current at the MCC of 40,000A rms symmetrical. The individual motor branch circuit feeds a 3-phase, 230V, 25-hp, squirrel-cage induction motor that has a 1.15 service factor.

Table 3.1 shows that the FLC rating for the 25-hp motor is 68A. To comply with the requirements for sizing circuit conductors, the conductors must be rated for at least 125% of FLC or 85A. From **Table 3.2,** it can be determined that selection of a No. 4 AWG, copper conductor with 75°C insulation would be acceptable. A motor controller is selected that has a minimum rating of 25 hp for the operating voltage, and the overload protective device required is set at no more than 125% of the motor nameplate FLC.

In **Fig. 3.8(A),** branch circuit short-circuit protection is provided using an inverse-time circuit breaker. From **Table 3.3,** it can be determined that the rating or setting for this CB can be at a maximum 250% of FLC. Because 170A (2.5 x 68A) does not correspond to a standard rating for CBs, the next higher standard rating, 175A, may be selected.

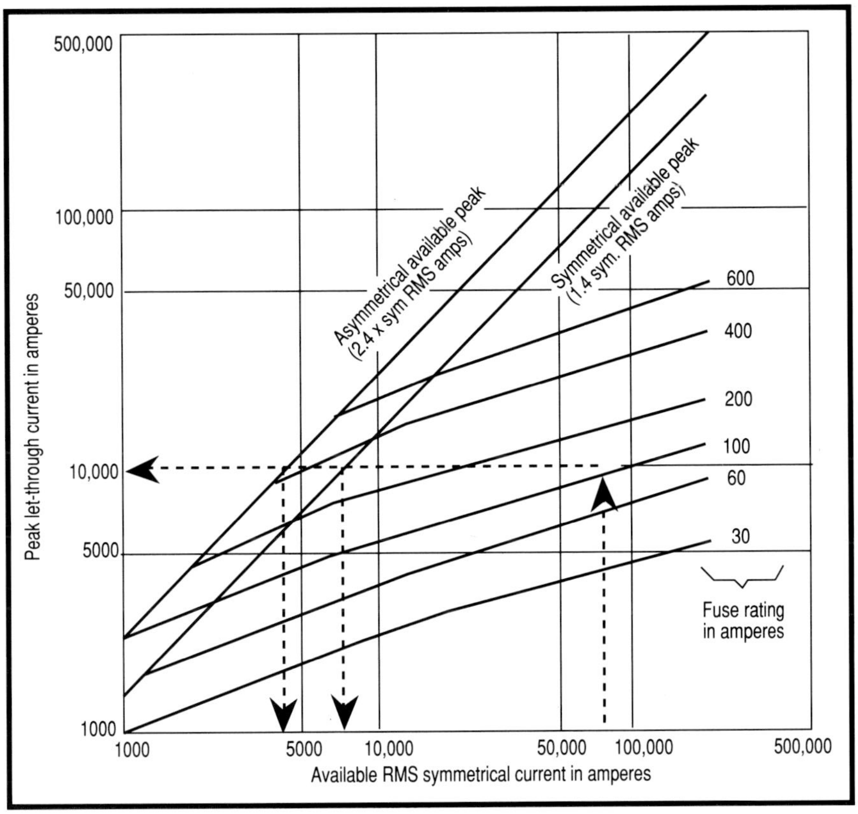

Fig. 3.6. Typical fuse peak let-through curves. In this example, with 80,000A rms symmetrical amps available, the fuse will allow a peak let-through of 10,000A. The fuse will begin to be current limiting (threshold current) somewhere in the range between 4500 and 7500A.

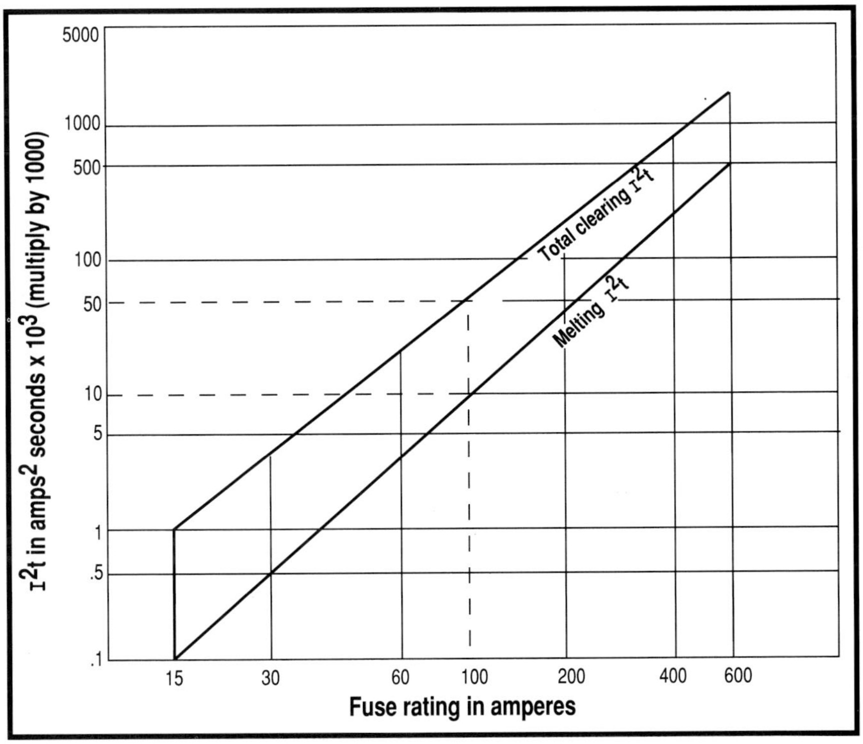

Fig. 3.7. Typical I^2t clearing and melting curves. In this example, the fuse will begin to melt at 10,000 and clear the fault after having let through 50,000 amp² sec of energy.

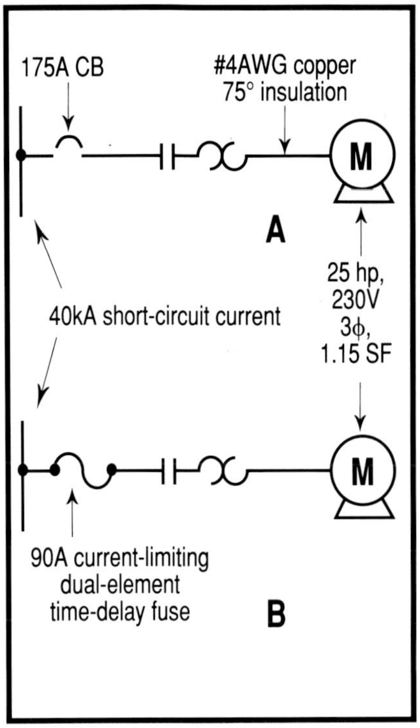

Fig. 3.8. Typical 3-phase, 230V, individual branch circuit. In (A) the circuit protective device is a 175A inverse-time circuit breaker. In (B), the device is a currrent-limiting, dual-element, Class RK1 time-delay fuse.

Induction Type Squirrel-Cage and Wound-Rotor Amperes				
HP	115V	230V	460V	575
1/2	4	2	1	.8
3/4	5.6	2.8	1.4	1.2
1	7.2	3.6	1.8	1.4
1 1/2	10.4	5.2	2.6	2.
2	13.6	6.8	3.4	2.
3		9.6	4.8	3
5		15.2	7.6	6.1
7 1/2		22	11	9
10		28	14	1
15		42	21	1
20		54	27	22
→ 25		**68**	34	27
30		80	40	3
40		104	52	4
50		130	65	

Table 3.1. Full-load current of 3-phase AC motors. A 230V, 25 hp motor has a FLC of 68A. (Derived from NEC Table 430-150).

Size	Temperature Rating		
	60°C (140 F)	75°C (167 F)	85° (185
AWG MCM	Type TW	Type FEPW	Ty
18			
16			18
14	20	20	25
12	25	25	3
10	30	35	4
8	40	50	[illegible]
6	55	65	70
4	70	**85**	95
3	85	100	1
2	95	115	1
1	110	130	[illegible]
1/0	125	150	[illegible]
2/0	145	175	19
3/0	165	200	21
4/0	195	230	2
250	215	255	2
300	240	285	
350	260	310	

Table 3.2. Ampacities of not more than three single insulated conductors, rated 2000V or less, in raceway in free air. A No. 4 AWG, 75°C-rated conductor has an ampacity of 85A. (Derived from NEC Table 310-16).

While this approach is typical and appears to comply with the NEC, it actually violates code rules and is potentially hazardous to personnel and equipment. To illustrate why, assume a one-cycle opening time for the CB. If a fault occurs on this motor's branch circuit conductors, 40,000A of available short-circuit current will flow through all circuit components to the point of the fault for at least one cycle.

The Insulated Cable Engineers Association Publication ICEA P-32-382 shows that a No. 4 AWG, copper conductor with 75°C-rated insulation has a short-circuit withstand rating of 17,100A. **Table 3.4** shows that the UL short-circuit withstand rating for the motor controllers is 5000A unless otherwise marked.

Whether "extensive damage" would occur to the circuit components under these circumstances is academic. There is no question that the conductors and controller in this case would be subjected to excessive thermal and magnetic stresses as a result of the fault. Good practice calls for the use of a current-limiting protective device that will limit the amount of available fault current to some value below the withstand rating of circuit components.

Assume the CB is replaced with a current-limiting, dual-element time-delay fuse with a UL classification of RK1 (very current limiting) and rated at some value less than the maximum permitted. Referring to **Table 3.3,** it can be determined that the fuse may be rated at a maximum of 175% of FLC (1.75 x 68A = 119A). However, in an effort to provide enhanced protection, a fuse rated at 125% of FLC or the next higher standard size, (90A) is selected.

This approach limits the available short-circuit current. A typical current-limiting fuse of this rating will limit the rms symmetrical current to 3700A, which is well below the 5000A maximum allowable at the controller. The circuit conductors are protected closer to their ampacity, 85A, and backup motor overload protection is also provided. A 100A fusible switch can be used to hold the 90A fuses.

A note of caution: In any application, it is necessary that a smaller-than-permissible-rating branch circuit protective device have a sufficient time delay to allow for motor starting current without opening the circuit. In addition, the smaller-sized device must have an interrupting-capacity rating at least equal to that of the available short-circuit current, and be capable of limiting the available short-circuit current, or be used where the available short-circuit current is otherwise limited.

Horsepower rating	RMS symmetrical amperes
1 or less	1000
1½ to 50	**5000**
51 to 200	10,000
201 to 400	18,000
401 to 600	30,000
601 to 900	42,000
901 to 1600	85,000

Table 3.4. UL regulations state that unless otherwise marked, motor controllers incorporating thermal cutouts or overload relays are considered suitable for use on circuits where the short-circuit current available is not greater than the values shown. A 25-hp-rated motor controller has a rating of 5000A rms symmetrical.

MCPs AND HIGH-EFFICIENCY MOTORS

Instantaneous, magnetic-only motor-circuit protectors (MCPs) provide excellent protection of motor circuits.

Type of Motor	Percent of Full-Load Curent			
	Nontime Delay Fuse	Dual Element (Time-Delay) Fuse	Instantaneous Trip Breaker	Inverse Time Breaker
Single-phase, all types				
No code letter	300	175	700	250
All ac single-phase and polyphase squirrel-cage and synchronous motors with full-voltage, resistor or reactor starting:				
No code letter	300	**175**	700	**250**
Code letter F to V	300	175	700	250
Code letter B to E	250	[illegible]	[illegible]	[illegible]

Table 3.3. Maximum rating or setting of motor branch-circuit short-circuit and ground-fault protective devices. An inverse-time breaker can be set to a maximum of 250% of the motor FLC. A dual-element time delay fuse is permitted to have a maximum rating of 175% of motor FLC. (Derived from NEC Table 430-152).

Hp	Standard design	High efficiency
1	27	28
1.5	36	38
2	46	48
3	59	62
5	86	90
7.5	121	128
10	156	164
15	226	240
20	288	305
25	369	393
30	447	478
40	606	651
50	773	834
60	932	1008
75	1173	1272
100	1575	1709
125	2015	2201
150	2464	2709
200	3375	3740

Table 3.5. Starting currents of typical standard and high-efficiency NEMA Design B, 1800 rpm, 460V, 3-phase motors. The difference is smaller for lower hp motors, but becomes more significant as size increases.

They respond only to excessive current, so they will trip within a cycle or two, depending on their setting. Because of this, the magnitudes of starting currents must be known so that units can be set within the limits required by the NE Code, while allowing for starting-current inrush.

Sensitivity to unusual motor circuit conditions can also be a problem. For instance, if the actual circuit parameters do not meet "standard" conditions, false tripping can occur. This condition has shown up more frequently when high-efficiency motors are being protected.

Energy-efficient motors, because of their low-loss design, frequently have a higher starting current than standard NEMA motors of the same hp and code letter. The nameplate code letter may be the same for both standard and energy-efficient motors; however, the peak portion of the starting current of an energy-efficient motor can be much higher than that of a standard motor. This affects the operation of the protective device.

In analyzing the difference between locked-rotor currents of standard vs. high-efficiency motors, the oscillogram in **Fig. 3.9** shows how the current varies as a function of time in one phase of a 3-phase machine during the first 40 cycles after being connected to the line. The first 30 cycles, while current is in general decline, is called the transient. The latter portion of the curve where the current is symmetrical at the zero line is called the steady state. The transient portion of the wave is sometimes referred to as the asymmetrical time.

A motor's kva code letter is determined in accordance with parameters that include values longer than the transient portion of the starting current. Thus, the nameplate code letter may be the same for both a standard and energy-efficient motor of the same rating. However, the transient portion (particularly the first cycle) of the starting current of an energy-efficient motor may be much higher than that of a standard motor. As seen in the oscillogram, the maximum value of the surge occurs at the end of the first half cycle. It is at this value that it must be determined whether or not an MCP will function properly and protect the motor without tripping.

Studies conducted have concluded that these problems result from the differences in winding resistances between the two types of motors, and is related to the locked-rotor power factor of the motor. Also, the peak value of the inrush seems to be a function of the size of the motor, and the peak value usually occurs during the first cycle of the starting current.

In **Table 3.5,** the peak values of inrush current are shown for 460V NEMA-rated motors for standard and high-efficiency configurations.

Fig. 3.10 shows locked-rotor power

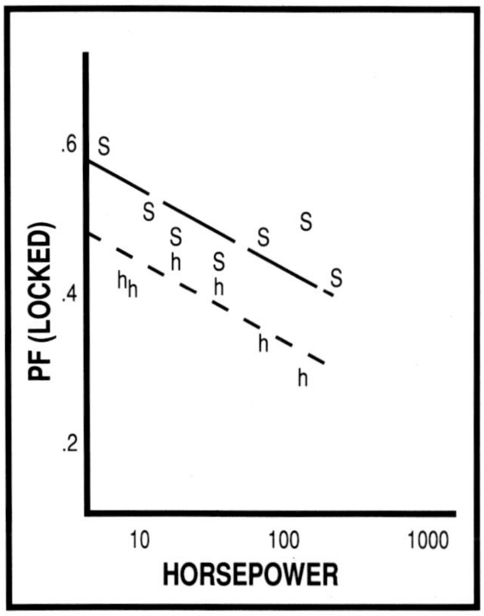

Fig. 3.10. Locked-rotor power factor plotted against hp. The larger the motor the lower the power factor, resulting in high peak transient current.

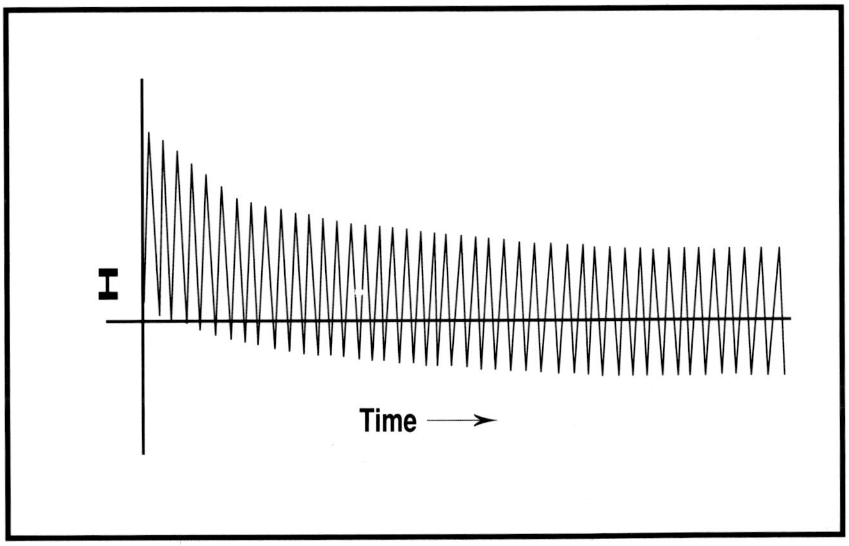

Fig. 3.9. Oscillogram shows how motor starting current varies as a function of time. At the left, current (I) is at a maximum and is termed "transient" until it reaches steady-state after 30 cycles.

factor on the vertical axis plotted against horsepower. The lines drawn through the points marked "S" indicate values for standard-efficiency motors, while points marked "H" represent values for high-efficiency motors. As motors become larger, the locked-rotor power factor decreases. The curves show that high-efficiency motors have lower locked-rotor power factors than standard motors. This is logical because high-efficiency motors have lower resistance than standard motors with resulting lower winding losses. An inescapable result of lower winding resistance is higher peak transient and a higher starting-current.

The conclusion to be drawn is that greater care must be taken in selecting and setting protective devices for modern motor circuits. To avoid nuisance trips when changing from a standard to high-efficiency motor, protective devices should be reset to handle the higher starting current.

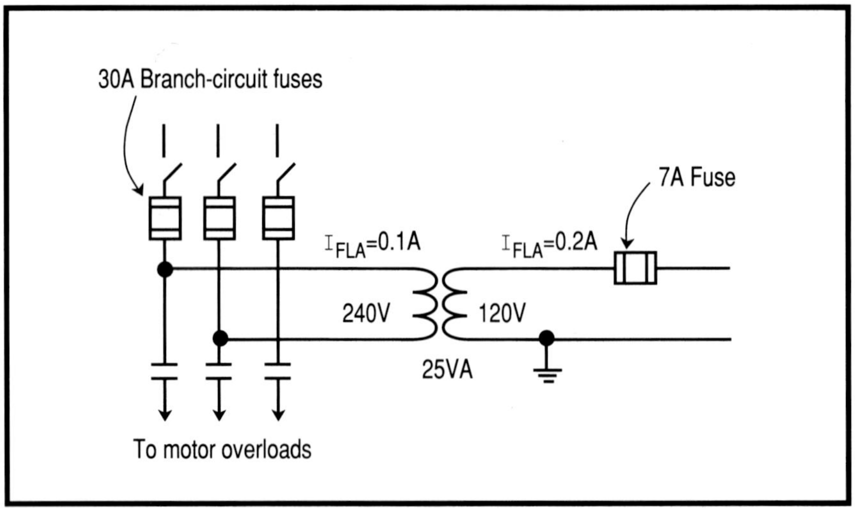

Fig. 3.11. With the rating of the transformer shown, primary circuit protection is not required. The fuse on the secondary protects the No. 18 AWG control-circuit conductors.

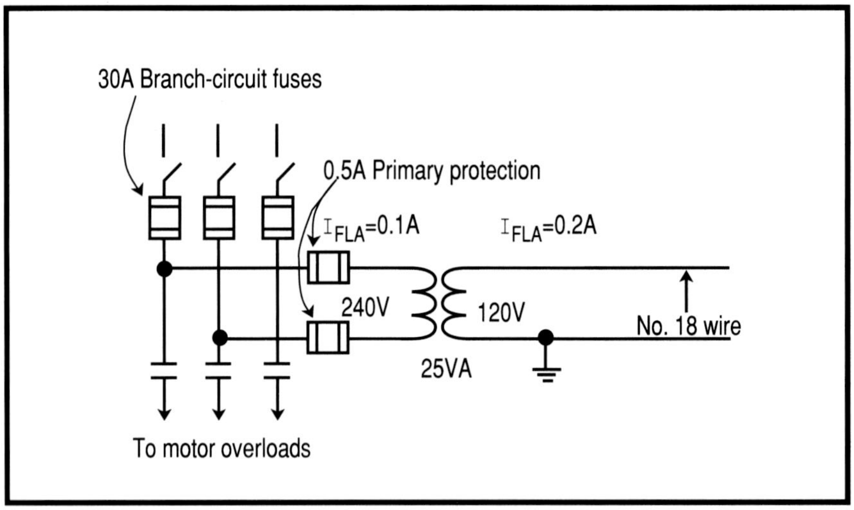

Fig. 3.12. The overcurrent device in the primary of the control-circuit transformer can be sized at 500% of primary FLC. A fuse is not required on the secondary because they are protected at 1A by the 0.5A fuses, well within the ampacity (6A) of the No. 18 conductors.

PROTECTING CONTROL TRANSFORMERS

An industrial control transformer is used to reduce the line voltage to the operating voltage of the motor's control circuit. Many of these transformers, even though installed in accordance with NE Code requirements, cause fires that damage electrical wiring and equipment.

The NEC is a minimum standard. It allows the rating of the protective device for the primary to be as much as 167% of the primary full-load current, for transformers whose primary current is less than 9A, and permits control-circuit transformers, whose primary full-load current is less than 2A, to have protective devices rated 500% of primary full-load current. Transformers rated at 50VA or less are not required to have overcurrent protection.

It might seem from these rules that smaller transformers are less likely to short-circuit or be overloaded. This is not the case. The reason for the flexibility is to prevent nuisance interruptions caused by the inrush current associated with control-circuit transformers, which can have peak values from 4 to 26 times the operating current of the transformers. As the rated size of the transformer decreases, the size of the fuse also will decrease. Smaller fuses have a reduced ability to withstand inrush current.

EXAMPLE: To fully appreciate the benefits of using overcurrent protection on control-circuit transformers, consider the transformer in **Fig. 3.11.** It is fed from the 30A branch circuit that supplies the motor to be controlled. Because of its low rating, overcurrent protection is not required for the primary of the transformer. If the primary windings of this transformer short-circuit, and the overcurrent that resulted was less than 30A, the response time of the 30A fuses would be such that extensive damage would be done to the control-circuit conductors, or the transformer could become hot enough to burst into flames. This is one of those situations where exceeding the requirements of the NEC is the most prudent action to take.

Consider the use of low-amperage fuses as an overcurrent device to protect the control transformer **(Fig. 3.12).** The primary full-load current is 0.1A. It is permissible in accordance with NEC rules to use overcurrent devices rated 500% of the primary full-load current.

This results in the selection of 0.5A standard fuses. Assume for this example that the inrush is 26 times the primary full-load current. Upon application of power to the branch circuit, the control transformer overcurrent device would be subjected to 2.6A peak current for the first half-cycle (.0083 sec).

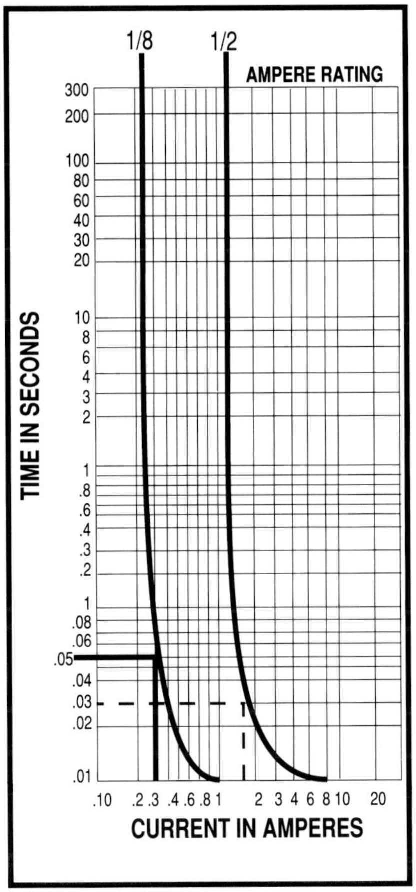

Fig. 3.13. Typical time-current characteristic curve of fast-acting fuses.

Time-current characteristic curves can show if the selected overcurrent devices can withstand the inrush for a half cycle. Because the ampere ratings on these curves are rms values, the 2.6A peak current must be converted to an rms value by dividing by 1.414. The calculated rms value in this example is about 1.83A.

From the manufacturer's time-current characteristic curve **(Fig 3.13),** these fuses can withstand about 1.83A rms for .03 sec, which should not result in nuisance interruptions on inrush. Thus, these devices will provide short-circuit protection for the control transformer, but an overload condition could still cause damage before the overcurrent protection opens the circuit.

The potential for damage due to overload can be minimized by reducing the rating of these fuses to some value less than 500% of primary full-load current. According to the code rules, primary overcurrent protection must not be rated more than 125% of primary full-load current. The primary full-load current is 0.1A x 1.25 = 0.125A (1/8 A). The characteristics curve shows that this fuse provides better overload protection. For example, if the primary current reached 0.3A, the fuse would blow in 0.05 sec, while the fuse rated at 0.5A would not clear an overload of twice that amperage. But the 1/8A fuse would not be able to withstand the 1.83 A inrush current. Because of the inrush, the overcurrent devices must be sized at 500% of primary full-load current at the sacrifice of better overload protection.

Cartridge fuses in disconnect switches used in motor controllers protect motor circuits from the effects of short circuits.

This does not have to be the case.

Although the inrush is greater than the operating current, it only lasts for ½ cycle, and therein lies the solution to providing functional short-circuit protection and a higher level of overload protection for the control transformer. The solution is time-delay fuses.

In **Fig. 3.14,** the full-load current is still 0.125A, or 1/8A. Because there is no time-delay fuse rated 1/8A, the next higher-rated fuse available, 15/100A (0.150A), and an inrush equal to 2.6A peak (1.83A rms), will be used.

The time-current characteristics curve **(Fig. 3.15)** shows that the fuse can withstand 1.83A rms for 2 sec and would not cause nuisance interruptions. Further, if during an overload the primary current reaches 0.3A rms, this fuse would clear the overload in 30 seconds, thereby reducing the risk of heat damage to the control-circuit con-

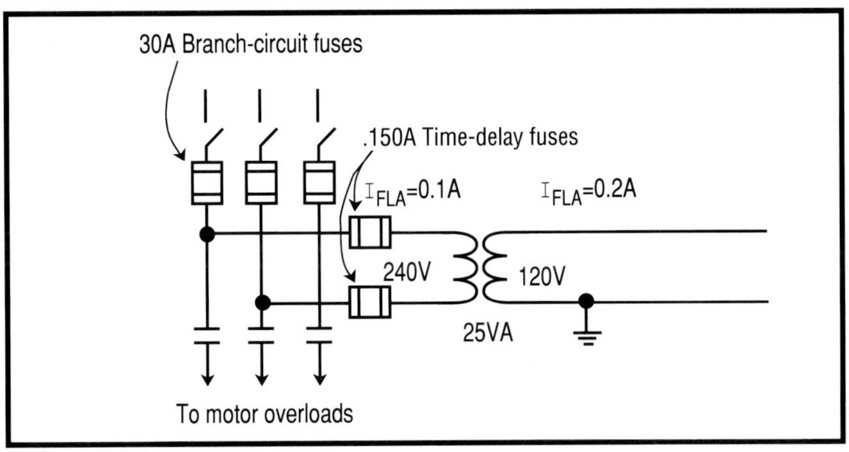

Fig. 3.14. The control-circuit transformer is protected by the time-delay fuses. Better short-circuit and overload protection is afforded.

ductors and control transformer.

For motor control-circuit transformers, the use of time-delay fuses, rated 125% of the primary full-load current, although not required by the NEC, is the most desirable approach to overcurrent protection. They will provide an extra measure of overload protection in addition to functional short-circuit interruption.

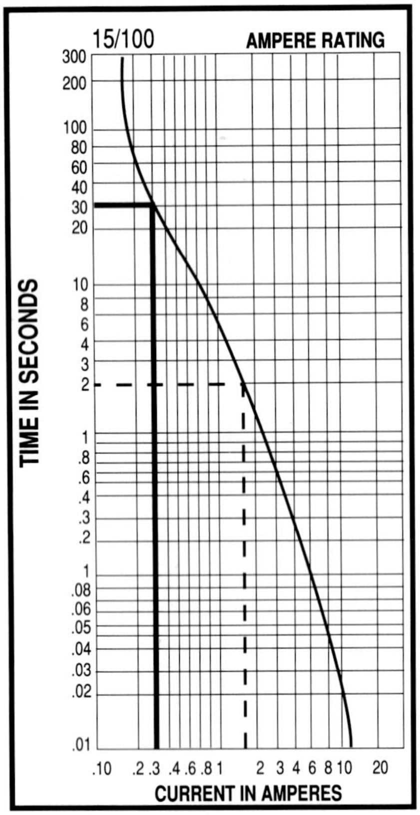

Fig. 3.15. Time-current characteristic curve of a typical 0.150A time-delay fuse.

Molded-case circuit breakers come in a wide variety of types to protect motor circuits. They include thermal-magnetic types, magnetic-only types, as well as combination types that include fuses that interrupt the circuit when fault current exceeds the rating of the circuit breaker.

4

MOTOR OVERLOAD PROTECTION

DAMAGE due to overloads accounts for most motor failures. An overload is any running current that is in excess of the rated operating current, up to and including locked-rotor current. An overload will manifest itself as an increase in the motor's operating temperature. This increase, if allowed to continue for a sufficient length of time, will cause damage to the motor by contributing to degradation of the insulating material used on the internal windings. A breakdown of this insulating material will ultimately result in motor failure. Fault currents due to shorts or grounds are not considered overloads.

Overloads can result from a variety of causes. These conditions include:

- attempting to drive a load that is greater than the horsepower rating for a given motor;
- heavy-inertia loads that require a large amount of torque for them to be put into motion;
- the motor being mechanically prevented from turning;
- line voltage less than approximately 80% of the motor's nameplate rating;
- excessive starting and stopping;
- failure in one phase of the power circuit (single phasing);

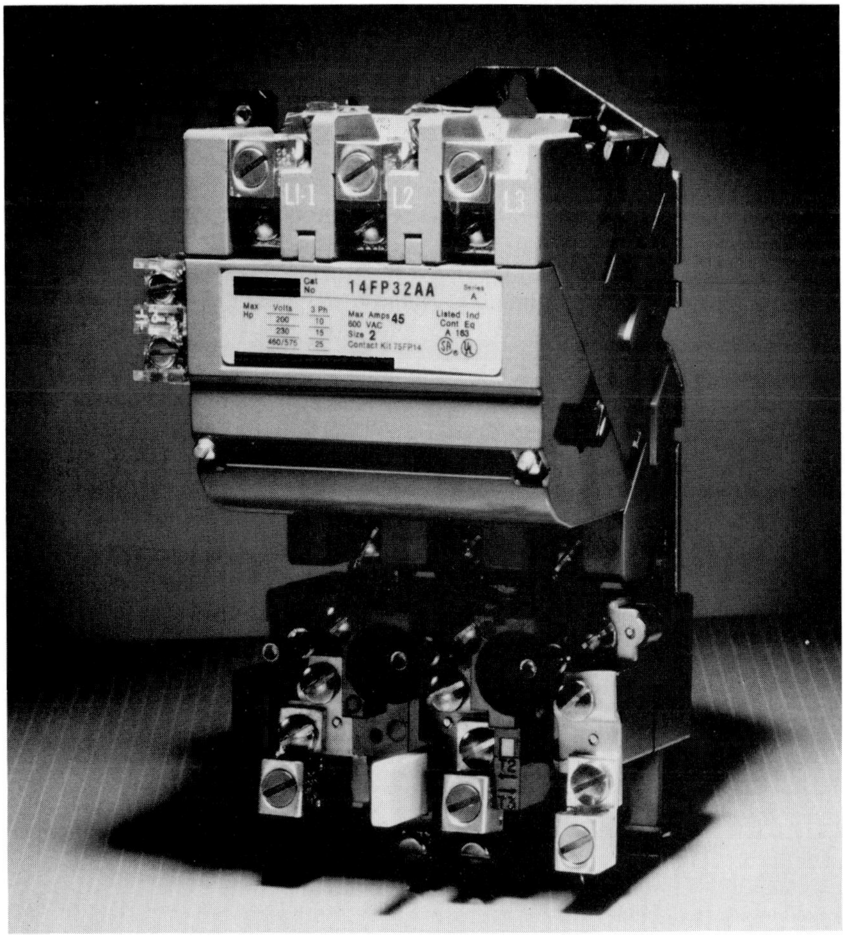

The overload relay mounted below the contactor of this NEMA Size 2 starter protects the motor by passing the current being drawn through heaters that cause an alloy to melt, or bimetal to deflect.

- an increase in frequency; and
- high ambient temperature.

OVERLOAD RELAYS

Fuses or circuit breakers may be used for overload protection, but they are not the best suited devices for protecting against these conditions in motors because of their inability to withstand the inrush or starting current, and excessive response time to moderate running overcurrents.

Protection against overloads can be achieved best by the application of overload relays in the motor circuit. They are made up of two components: a sensing device and a tripping mechanism. The sensing devices detect a rise in heat or current. Current-sensing devices react directly to the current being drawn, while heat-sensing devices react indirectly to the current.

Current-sensing overload relays usually employ current transformers (CTs) that provide an output proportional to the current being drawn at any time during the motor's operation. This type of sensing is used typically in instances where operating current is in excess of approximately 240A (NEMA Size 5 starters). CTs also are used with microprocessor-based, multifunction, electronic motor-protection units.

Heat-sensing overload relays use the heat generated by the motor current over a period of time (I^2t) as the basis for their operation. The two most commonly used thermal sensing techniques are the eutectic-alloy (solder-pot), thermal overload relay, **Fig. 4.1**, and the bimetal thermal overload relay, **Fig. 4.2**.

Eutectic alloy is a form of metal having a sharp melting point. It changes quickly from a solid to a liquid. Common solder is an example of a eutectic alloy. This trait allows the relay to operate at a very specific temperature with a minimal transition time from contacts open to contacts closed, or vice versa.

The motor's current flows through a resistive wire "heater" that is in close physical proximity to the eutectic alloy. The heater is selected based upon motor's FLA requirements. Heat generated by the current flowing through the heater will not be enough to melt the solder as long as the load remains at a level below the rating or setting of the overload relay. If the load increases to excessive levels, current increases proportionally. When enough heat is generated in the heater by excess current, the eutectic alloy will melt, caus-

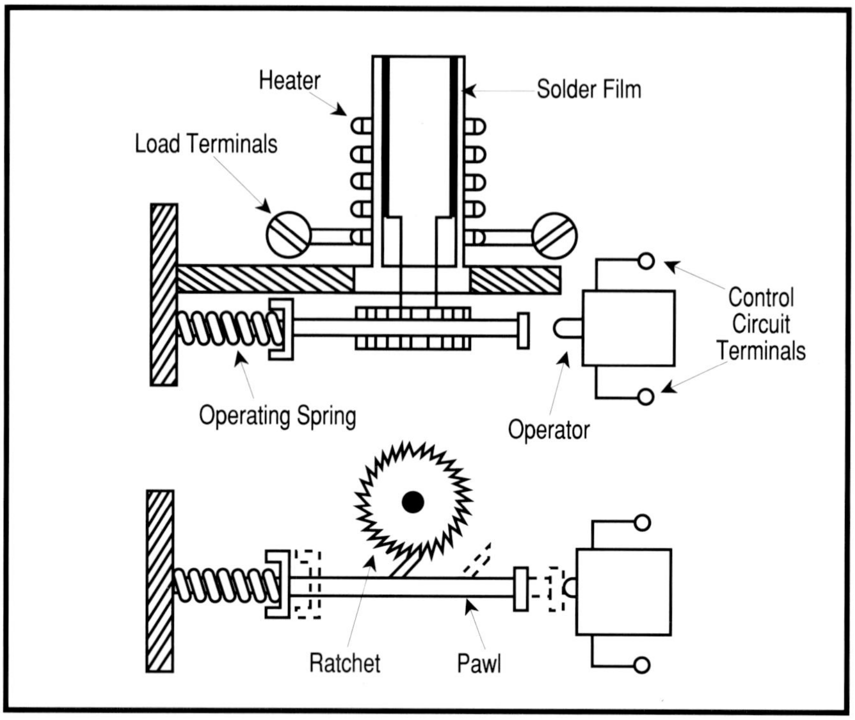

Fig. 4.1. Eutectic-alloy (solder-pot) thermal overload relay. Current in the motor's power circuit flows through a heater, which is wound around a small cylinder that contains the eutectic alloy. A shaft, with a good part of its length embedded in the alloy, has a ratchet wheel attached to its end. A spring-loaded actuator is held in position by a catch that engages the ratchet wheel. When the temperature rises in the heater as a result of excess current drawn during an overload, the solder melts. The ratchet is now free to turn and no longer can restrain the spring-loaded actuator, which moves and opens a contact in the starter's control circuit. Deenergizing the control circuit causes the main motor contacts in the power circuit to open, deenergizing the motor. Once the eutectic alloy has resolidified, the relay is manually reset by moving the actuator to the position in which it recompresses the spring.

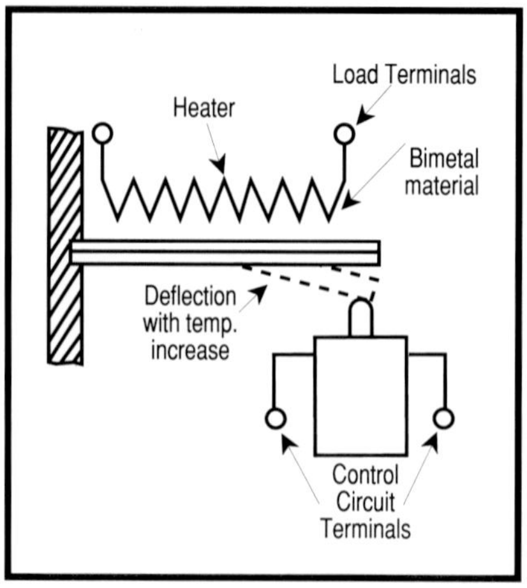

Fig. 4.2. A basic bimetal thermal-sensing overload relay.

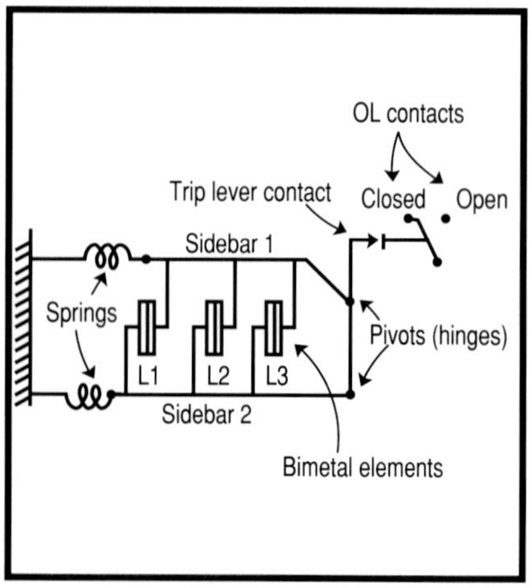

Fig. 4.3. A slide-bar bimetal thermally-operated overload relay.

ing the relay to operate and open contacts in the starter's control circuit.

Bimetallic elements utilize the difference in the expansion coefficients of two different metals that are welded together back-to-back to form a bimetal strip. A resistive wire heater in the motor starter power circuit is heated by the current flow and the bimetal strip reacts proportionally. When a sustained value of excess current flows, the resulting heat causes the bimetal to physically bend. This mechanical action is used to actuate contacts that interrupts power to the starter's control circuit. Unlike the eutectic type, the bimetal thermal overload relay can be automatically or manually reset.

It might appear that problems with nuisance tripping could occur as a result of inrush current during startup. This is not the case. The inherent time delay in the mechanical action of the thermally-operated overload relays prevents inrush currents from causing nuisance tripping.

While use of either the eutectic-alloy or bimetal-strip sensing will protect the motor from most overloads, they have only a limited ability to prevent motor damage during single phasing, even with all three phases being monitored. For that reason, some bimetal thermally-operated overload relays **(Fig. 4.3)** incorporate an additional feature. A slide-bar allows this type of overload relay to use both the action of the additional heating that will be experienced by the two remaining hot legs, as well as the reaction due to cooling in the open leg, to provide enhanced protection against damage due to single phasing.

APPLYING THERMAL OVERLOAD RELAYS

Thermal overload relays are identified by Class designations that are based on the amount of time an OL relay requires to interrupt a current equal to 600% of its normal current rating. For example, a Class 10 overload relay will trip in 10 sec at six times its rated current. A Class 20 will trip in 20 sec, and a Class 30 will trip in 30 sec. The class designation is an important consideration when applying OL relays in motor-control circuits.

The current rating of OL relays are expressed in amps and are based on a standard ambient temperature of 40°C.

To properly protect a motor against overloads certain motor operating information must be determined: full-load current (FLC), and service factor or temperature rise. These operating parameters can be found on the motor's nameplate. Do not confuse the nameplate ambient-temperature rating with temperature rise.

The following are examples of typical applications that clarify how OLs are applied.

EXAMPLE 1. 50-hp, 230V, wound-rotor crane motor: FLC= 120A; service factor= 1.15. Rules spelled out in the code apply.

SOLUTION. OL protection for continuous-duty motors of more than 1 hp, with a marked service factor not less than 1.15, or marked temperature rise not over 40°C, must be rated or set to trip at not higher than 125% of the motor's FLC. The maximum rating or setting for the OL relay would be 150A (1.25 x 120A).

EXAMPLE 2. 10-hp, 460V, squirrel-cage industrial mixer motor: FLC= 13.5A; service factor= 1.0.

SOLUTION. Overload protection for continuous-duty motors of more than 1 hp, with a marked service factor less than 1.15, or a marked temperature rise over 40°C must be rated or set to trip at not higher than 115% of the motor's full load current. The calculated value is 15.5A (1.15 x 13.5A).

The code allows for upsizing an OL relay when the selected relay does not permit the motor to start or carry the load. The setting or rating may be increased from 125% to a maximum of 140% of the motor's FLC for motors with a service factor greater than 1.15, and 130% for motors with other than a 1.15 service factor.

EXAMPLE 3. Unlike general-purpose motors, OL protection for hermetic (sealed) compressor motors is based solely on the motor's FLC. Service factor and temperature rise are irrelevant. Further, the maximum multiplier used to calculate the value for the OL protection by the OL relay is 140% instead of 125%. NEC Article 440 permits the use of this higher multiplier because the presence of the refrigerant makes these motors inherently cooler during operation and allows them to be safely operated at slightly elevated currents. For hermetic compressor motors, the 140% multiplier is a maximum. Upsizing of the OL relay is not permitted.

Heater Type No.	Full Load Amps.							
	Size 00	Size 0	Size 1	Size 1P	Size 2	Size 3	Size 4	Size 5
W10	0.19	0.19	0.19					
W11	0.21	0.21	0.21					
W12	0.23	0.23	0.23					
W13	0.25	0.25	0.25					
W14	0.28	0.28	0.28					
W15	0.31	0.31	0.31					
		[illegible]	4.08					
W43	4.52	4.52	4.52					226
W44	4.98	4.98	4.98					249
W45	5.51	5.51	5.51		5.80			276
W46	6.07	6.07	6.07					
W47	6.68	6.68	[illegible]					

Table 4.1. Part of a typical NEMA-type starter overload heater selection chart.

OL HEATERS

Current ratings for nonadjustable, adjustable, and interchangeable OL relay heater elements are expressed in different ways, depending upon the manufacturer. Either the actual maximum current rating of the relay heater is used, or the maximum motor full-load current is given. The first method directly indicates the ampere rating of the device, while the second lists the device not by its current rating, but rather by the full-load current of the motor it will properly protect.

The most familiar method (NEMA) of matching an OL relay to a motor involves selecting an interchangeable heater element, which is rated at the desired current value (see **Table 4.1**). The other method (IEC) requires selecting an OL relay, which is adjustable over a specified range of current, and setting the adjustment dial to the desired current value (see **Table 4.2**).

Regardless of whether an interchangeable element or adjustable overload is utilized, different methods are used by different manufacturers to compile their selection tables. Most selection tables list these devices according to motor FLC. When an OL heater element is rated this way, the calculations required by the NEC to determine the necessary level of protection have already been completed. Typically, it is assumed that the motor has a service factor of 1.15 or greater and a temperature rise not over 40°C, which allows the motor to be protected up to 125% of the nameplate FLC rating.

For example, a device rated at 10A in the selection table is intended for use with a motor that has a 10A FLC. This 10A device will cause the circuit to be interrupted at a current value greater than 10A, but not greater than 12.5A (1.25 x 10A). NEMA standards permit classifying OL heater elements in this manner, but require the manufacturer to provide conversion factors for selecting devices to protect motors that have a service factor less than 1.15 or a temperature rise over 40°C (see **Table 4.3**).

Another consideration for selecting OL relay heater elements is the ambient operating temperature of the motor and the OL relay. Guidelines for selecting the proper rated device when the motor and OL relay are in different ambients are in the manufacturers' catalogs. This is only a concern for nonambient-compensated relays.

MOTOR PROTECTIVE RELAYS

Microprocessor-based motor protective relays (MPR) do the work of a dozen separate protective relays. Microprocessor relays provide better monitoring, increased accuracy, simplicity of application, and lower overall costs, particularly where a variety of functions is needed. One MPR can replace numerous individual protective relays.

For example, for 3-phase AC induction and synchronous motors, MPRs can provide instantaneous overcurrent, overload, locked rotor, jam, instantaneous ground overcurrent, phase differential current, phase unbalance,

Overload Protection Modules			
Thermal Setting Range (Amps)	Magnetic Setting Range (Amps)	Standard Module w/ Thermal & Magnetic Trip	Magnetic Only Module (Us Ranges in COl. 2 Only)
32A Models			
0.25–0.40	2.4–4.8	LB1-LC03M03	N/A
0.40–0.63	3.8–7.6	LB1-LC03M04	N/A
0.63–1.0	6.0–12	LB1-LC03M05	N/A
1.0–1.6	9.5–19	LB1-LC03M06	LB6-LC03M06
1.6–2.5	15–30	LB1-LC03M07	LB6-LC03M07
2.5–4.0	24–48	LB1-LC03M08	LB6-LC03M08
4.0–6.3	38–76	LB1-LC03M10	LB6-LC03M10
6.3–10	60–120	LB1-LC03M13	LB6-LC03M1[illegible]
10–16	95–190	LB1-LC03M17	LB6-LC03M[illegible]
16–25	150–300	LB1-LC03M22	[illegible]
23–32	[illegible]	[illegible]	[illegible]

Table 4.2. Part of a typical IEC-type starter overload protection selection chart.

MOTORS RATED FOR CONTINUOUS DUTY:

MOTORS WITH MARKED SERVICE FACTOR OF NOT LESS THAN 1.15, OR MOTORS WITH A MARKED TEMPERATURE RISE NOT OVER 40°C.

1. **The Same Temperature at the Controller and the Motor** – Select the "Heater Type No." with the listed "Full Load Amps." nearest the full load value shown on the motor nameplate. This will provide integral horsepower motors with protection between 110 and 120% of the nameplate full load currents.
2. **Higher Temperature at the Controller than at the Motor** – If the full load current value shown on the motor nameplate is between the listed "Full Load Amps.", select the "Heater Type No." with the higher value. This will provide integral horsepower motors with protection between 115 and 125% of the nameplate full load currents.
3. **Lower Temperatures at the Controller than at the Motor** – If the full load current value shown on the motor nameplate is between the listed "Full Load Amps.", select the "Heater Type No." with the lower value. This will provide integral horsepower motors with protection between 105 and 115% of the nameplate full load currents.

ALL OTHER MOTORS RATED FOR CONTINUOUS DUTY (INCLUDES MOTORS WITH MARKED SERVICE FACTOR OF 1.0)

Select the "Heater Type No." one rating smaller than determined by the rules in paragraphs 1, 2 and 3. This will provide protection at current levels 10% lower than indicated above.

*Rules 2 and 3 apply when the temperature difference does not exceed 10°C (18°F)

Table 4.3 NEMA requires that manufacturers provide conversion factors for overload heater selection.

phase reversal, loss of load, motor winding overtemperature, motor bearing overtemperature, load bearing overtemperature, and others. In addition, they provide a digital display of motor conditions, status of trip output and alarm relays, trip and time-delay values, and field-settable data to define the specific motor and system being protected (CT ratios, full-load amperes, type and number of RTDs, etc.). An emergency restart is also available to allow restarting of a hot motor. MPRs primarily are used to protect induction motors rated 200 hp and higher.

The brain of the relay is a microprocessor that organizes and processes all input and output information, reducing the data to a digital form for easy processing. It samples all data (points) virtually instantaneously. For example, to sense motor currents and temperatures, the MPR receives simultaneous inputs from phase CTs, ground CTs, phase differential CTs, and from RTDs. Self-checking tests run continually to assure that everything is working properly.

MPRs use four basic processes to provide motor protection.

Sampling. As with any multiplexing system (sequential scanning of discrete points), the accuracy and rate of scanning are important to assure that all dynamic data is up-to-the-instant. A microprocessor assures that no data is older than a few milliseconds. For vital data, a "speed-up" circuit is used to initiate a trip on a predetermined temperature rate-of-rise.

Modeling. Thermal modeling techniques duplicate the motor's damage curve under all performance conditions. For example, a simple bimetallic thermal overload relay uses a heater element sensitive to motor current in an attempt to duplicate the real heating curve of the motor. This unsophisticated "model" neglects many factors affecting motor heating, but nonetheless it is quite suitable for many noncritical applications with smaller motors. For larger motors and critical applications, certain important dynamic factors cannot be ignored. The MPR uses a model based on symmetrical components, including the effects of any voltage unbalance of applied phase voltages. Without taking into account this very common condition, a small percentage of voltage unbalance can produce severe errors in rotor heating protection. The microprocessor automatically solves the associated symmetrical component mathematical equations to provide a more faithful and accurate model of the total motor damage curve.

Freeze on trip. Under normal motor operating conditions, current input signal data is read and stored in the relay. Whenever tripping occurs, all data is retained until such time as a reset command is given. Until that time, the "frozen" signal data can be read and/or logged to help analyze what caused the fault to occur. Furthermore, an electrically erasable programmable read-only memory (EEPROM) chip retains all initial value settings plus the trip-out values and LED signal, even if all power input is lost. This avoids entering all initial relay settings again, saving time and effort.

Self diagnostics. MPRs are furnished with alarm LEDs and other trip indications, which can speed up location and correction of an internal problem that occurs. These features provide a valuable tool that speeds troubleshooting.

In addition to functions already mentioned, data that can be integrated into the motor-protection scheme includes: values of each phase-load current, numerous winding temperature points, bearing temperatures, ground-fault currents, and many others. However, because all data function requirements are arranged into three logical groups, the MPR becomes quite easy to use.

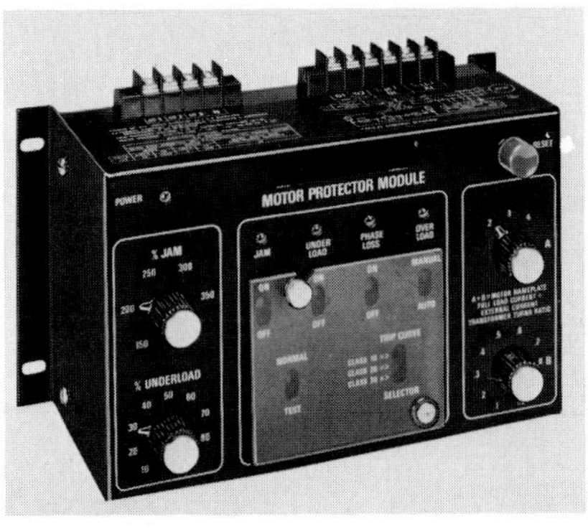

Solid-state devices are now available that protect critical motors not only from overload, but also phase loss, underloads, and various other events with great accuracy.

The first group of functions spells out the specific hardware system that provides information for the microprocessor to make its numerical calculations. As an example, if an RTD is used, a special circuit provides calibration and automatic compensation by removing the effect of lead resistances. Furthermore, the type and quantity of RTDs can be identified so that ordering of replacements is simplified.

The next group of functions are the data settings that describe the protection curve selected for the specific motor. This process is almost the same as the selection for conventional relays.

The final group of functions provides the information for the type of annunciation and alarm procedures desired. During all of this data handling, the microprocessor in the MPR continually sequences through all of its setpoints, processes the input signals, compares the actual values to limit values, does all of the scaling calculations to yield convenient units, performs myriad calculations for the symmetrical components modeling system, keeps track of the thermal history of the motor, provides the required outputs for alarms and tripping, and additional functions.

CHOOSING A PROTECTION SYSTEM

A decision on whether standard techniques of overload protection should be used or MRPs applied depends upon engineering and economic considerations. However, justification of any protection system must take into account some basic principles.

Safety. Personnel safety must always be a foremost consideration.

Cost of the motor. The size and expense of large motors, some reaching 1000 hp or more, will certainly justify sophisticated protection to avoid damage to this costly investment.

Cost of downtime. Downtime of a continuous process operation, even those not involving particularly large or expensive motors, could involve lost production and a high cost for cleanup and restarting. In cases like this, an MPR with a "thermal memory" and emergency restart feature, could permit an overload condition to continue, thus postponing shutdown to a less critical time. This protection takes advantage of actual motor temperature measurements. Without this protection, the motor would trip at its nominal overload rating, which must assume maximum ambient and full-load operating temperatures.

Maintenance considerations. The power of the microprocessor has reduced its maintenance to a few simple procedures. Recommended preventive maintenance consists of making an annual verification of the initial acceptance test. This test is performed in the field with a minimum of equipment. Its purpose is to assure that the MPR is still performing as originally set up. All calibration is accomplished by a single level-adjusting potentiometer, with the microprocessor completing all other adjustments. Troubleshooting and repair work are simplified and require only the direct replacement of the few printed circuit boards that make up the MPR. Depending on the number of MPR units used at a given location, a spare MPR or at least a spare microprocessor board could easily be justified.

OTHER TECHNIQUES

For large motors, it is impractical to pass the high phase current directly through the heater element, so a current transformer (CT) is used. The lower CT secondary current is proportional to the motor line current and is applied to the heater.

For medium-voltage motors, CTs are also used, with the secondary current applied to the heater of a thermal OL, or connected to a switchgear-type induction-disk or solid-state motor relay. Individual ground-fault protective relays can also be added to the motor circuit if desired.

Many large low-voltage motors and medium-voltage motors have temperature sensors (thermocouples or RTDs) embedded in their windings at key points and sometimes also in the bearings. When one of these detects excessive temperature, it actuates a relay that opens the coil circuit of the starter, disconnecting the motor.

The same technique of embedded detectors is also used on some small motors. Thermal protection is provided directly, rather than through heater-actuated relays. Built-in thermal switches or thermistors are used that on motor overtemperature open the circuit to the motor either directly or through a control unit. Integral sensors, however, present a problem for maintenance because the motor must be disassembled to replace them if they fail.

CT's are used to reduce the current that will appear in measuring circuits.

Medium voltage motors require special relaying schemes that help protect them from damage.

5

DESIGNING MOTOR AND CONTROL CIRCUITS

UNDERSTANDING the rules spelled out in the National Electrical Code (NEC) is critical to the proper design of motor circuits and the control circuits that start, stop, and otherwise regulate the operation of motors.

APPLICABLE NEC ARTICLES

Two articles in the NEC are directed specifically to motor applications:

Article 430, "Motors, Motor Circuits, and Controllers" covers application and installation of motor circuits and motor control hookups, including conductors, short-circuit and ground-fault protection, starters, disconnects, and overload protection.

Article 440, "Air Conditioning and Refrigerating Equipment," contains provisions for motor-driven equipment and for branch circuits and controllers for the equipment. It also takes into account the special considerations involved with sealed (hermetic-type) motor compressors, in which the motor operates under the cooling effect of the refrigeration.

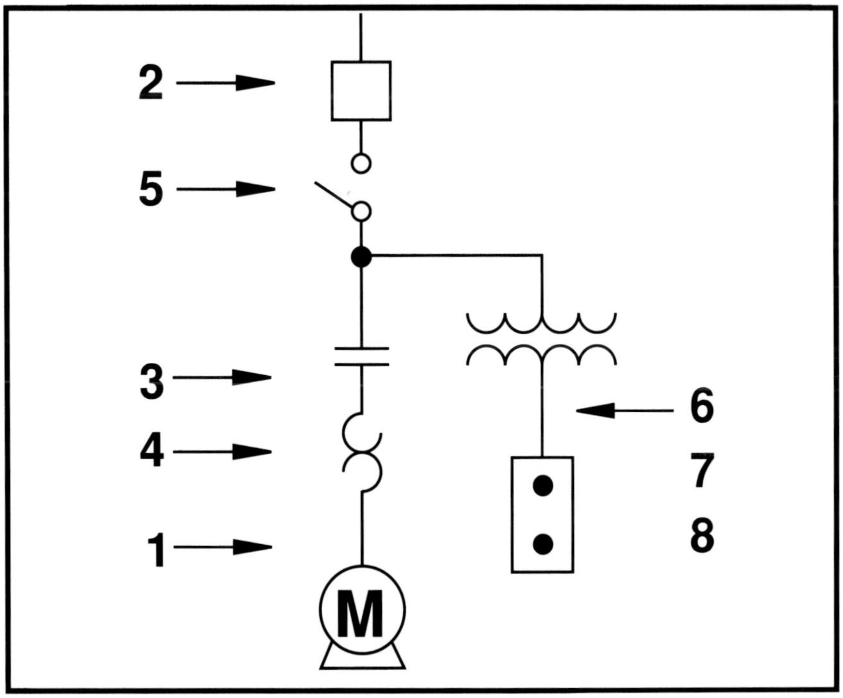

Fig.5.1. Follow these steps in numerical order.

The rules of Article 440 are in addition to, or are ammendments of, the rules given in Article 430 for motors in general. The basic rules of Article 430 also apply to air-conditioning and refrigerating equipment unless exceptions are indicated in Article 440. Article 440 further clarifies the application of NE Code rules to this type of equipment.

HVAC equipment that does not incorporate a sealed (hermetic-type) motor compressor must satisfy the rules of Article 422 (Appliances), Article 424 (Space Heating Equipment), or Article 430 (conventional motors), whichever apply. For instance, where refrigeration compressors are driven by conventional motors, the motors and controls are subject to Article 430, not Article 440.

Furnaces with air-conditioning evaporator coils installed must satisfy Article 424. Other equipment in which the motor is not a sealed compressor and which must be covered by Articles 422, 424, or 430 includes fan-coil units, remote commercial refrigerators, and similar equipment.

Room air conditioners are covered in Part G of Article 440, but must also comply with the rules of Article 422.

Household refrigerators and freezers, drinking-water coolers and beverage dispensers are considered by the code to be appliances, and their application must comply with Article 422 and must also satisfy the rules of Article 440, because such devices contain sealed motor-compressors.

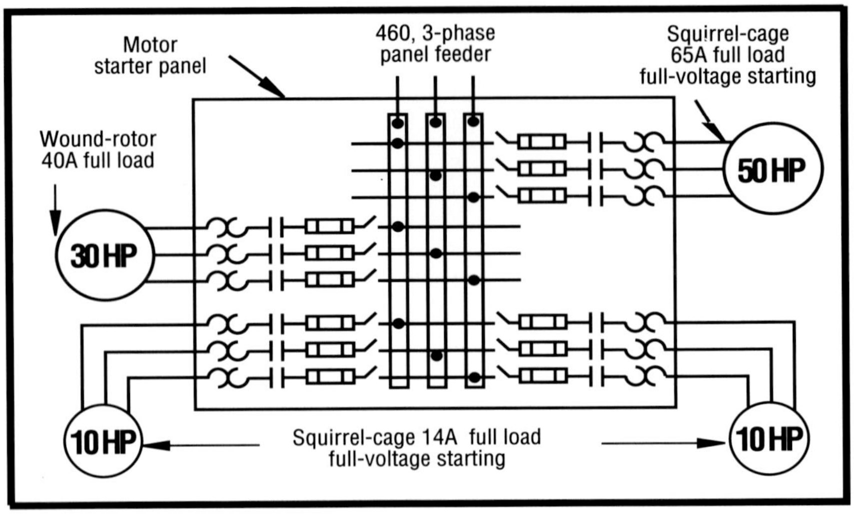

Fig. 5.2. Sizing branch-circuit and feeder conductors.

DESIGN CONSIDERATIONS

The eight basic elements that the code requires the designer to account for in any motor circuit is shown in **Fig. 5.1**. Although these elements are shown separately here, there are certain cases where the code will permit a single device to serve more than one function. For instance, in some cases, one switch can serve as both disconnecting means and controller. In other cases, short-circuit protection and overload protection can be combined in a single circuit breaker (CB) or set of fuses. Basic code requirements concerning these elements will be considered in the order shown in numerical order.

1. SELECTING CIRCUIT CONDUCTORS

The basic code rule says that the conductors supplying a single-speed motor used for continuous duty must have a current-carrying capacity of not less than 125% of the motor full-load current (FLC) rating. In the case of a multi-speed motor, the selection of branch-circuit conductors on the line side of the controller must be based on the highest of the FLC ratings shown on the motor nameplate.

It should be noted that all calculated conductor sizes are minimum, being based on temperature rise only. The calculations do not take into account voltage drop or power loss in the conductors. Such considerations frequently require increasing the size of branch-circuit conductors.

A typical circuit feeding four different motors from a panel is shown in **Fig. 5.2**. It is necessary to refer to the table in Article 430 that lists the FLC for the motor in question.

Feeder conductors supplying two or more motors must have a current rating of not less than 125% of the full-load current rating of the largest motor supplied, plus the sum of the full-load current ratings of the other motors supplied. In the example, assuming a 65A FLC for the 50-hp motor; 40A for the 30-hp motor; and 14A for the 10-hp motors; the panel feeder is sized:

$$(1.25)(65) + 40 + 14 + 14 = 149.25\text{A}.$$

Sizing branch-circuit conductors. For continuous-duty motors, the ampacity of branch-circuit conductors supplying a single motor must not be less than 125% of the motor full-load current rating. For the 50-hp motor, the branch circuit conductors must have a minimum ampacity of:

$$1.25 \times 65 = 82\text{A}.$$

However, motors used for short-time, intermittent, periodic or other noncontinuous duty impose varying heat loads on conductors. In such cases the conductors from the controller to the motor must have an ampacity not less than the percentage of the motor nameplate current rating shown in Article 430 table on duty-cycle service. Conductor sizing, therefore, varies with the application. But, any motor is considered to be for continuous duty unless the nature of the apparatus that it drives is such that the motor cannot operate continuously with load under any condition of use.

A wound-rotor motor is treated by the NEC in much the same way as squirrel-cage induction motors. The motor primary current also is listed in the FLC table. The values given in the table must be multiplied by a factor of 1.25 to determine the ampacity requirements of the conductors.

Conductors connecting the secondary of a wound-rotor induction motor to the controller must have a current-carrying capacity at least equal to 125% of the motor's full-load secondary current if the motor is used for continuous duty.

Conductors from the controller of a wound-rotor induction motor to its resistors, if they are external, must have an ampacity in accordance with the table for secondary conductors.

Motor current rating. The code has definite provisions for determining a motor's FLC.

For general motor applications (excluding applications of torque motors and sealed hermetic-type refrigeration compressor motors), whenever the current rating of a motor is used to determine the current-carrying capacity of conductors, switches, fuses, or circuit breakers, the values given in the FLC table for the appropriate motor should be used instead of the actual current rating marked on the motor nameplate. Overload protection, however, is based on the marked motor nameplate rating.

Conductors from a DC motor controller to separately-mounted power accelerating and dynamic braking resistors in the armature circuit must be sized per the table of conductor ampacity in percent of motor FLC. If an armature shunt resistor is used, the power accelerating resistor conductor ampacity must be calculated using the total of motor FLC and armature shunt resistor current. The shunt resistor conductors must have an ampacity of not less than that calculated from the percent of motor FLC listed in the table. The rated shunt resistor current is to be used as the FLC.

Rules for torque motors, shaded-pole motors, permanent split-capacitor motors and AC adjustable-voltage motors are also given in Article 430.

Article 440 covers sealed (hermetic-type) refrigeration compressor motors. The actual nameplate FLC of the motor must be used in determining the current rating of the disconnecting means, the controller, branch-circuit conductors, short-circuit and ground-fault protective devices, and motor overload protection. When such equipment is marked with a branch-circuit selection current, it must be used instead of the rated load current to determine the rating or ampacity of the conductors.

2. PROTECTING AGAINST SHORTS AND GROUNDS

The NEC requires that branch-circuit protection for motor circuits must protect the circuit conductors, the control apparatus, and the motor itself against overcurrent due to short circuits or ground faults.

Branch-circuit protective device. The protective device (circuit breaker of fuses) for an individual branch circuit to a motor must be capable of carrying the starting current of the motor without opening the circuit.

The code then places maximum values on the ratings or settings of such overcurrent devices. Such devices must not be rated in excess of the values given in Article 430 table of full-load currents for the appropriate motor. In case the values do not correspond to the standard sizes of fuses, to the ratings of nonadjustable CBs, or to possible settings of adjustable CBs, and the next lower value is not adequate to carry the load, the next higher size, rating or setting may be used.

In the **Fig. 5.2** example, the rating of nontime-delay fuses to protect the 50-hp motor is:

$$65 \times 3 = 195A$$

Because it is not a standard size, 200A fuses may be used if 175A fuses are not adequate.

Where absolutely necessary for motor starting, Article 430 permits higher setting up to a perscribed maximum.

For a multispeed motor, a single short-circuit and ground-fault protective device may be used for one or more windings of the motor, provided the rating of the protective device does not exceed the applicable percentage of the nameplate rating of the smallest winding protected. As an alternate, a single short-circuit protective device sized for the FLC of the highest-current winding may be used if each winding is provided with an individual overload device sized for the FLC of the individual winding, and the branch-circuit conductors are sized for the FLC of the highest full-load current winding.

The code establishes maximum values for branch-circuit protection, setting limits for safe applications. However, use of lower rated branch-circuit protective devices is obviously permitted by the code and offers opportunity for economies in selection of circuit breakers, fuses and the switches used with them, panelboards, etc. It is only necessary that the branch-circuit device that is smaller than the maximum permitted rating must have sufficient time delay in its operation to permit the motor starting current to flow without opening the circuit. But a circuit breaker for branch-circuit protection must have a continuous current rating of not less than 115% of the motor full-load current.

Where maximum protective device ratings are shown in a marked controller or are otherwise provided with the equipment, they must not be exceeded even if higher values are allowed by the NEC rules.

Fully-magnetic CBs. The NEC recognizes the use of an instantaneous-trip circuit breaker (without time delay) for short-circuit protection of motor circuits. Such breakers (also called "magnetic-only" breakers) are acceptable only if they are adjustable and are used in combination starters approved for the purpose, and have separate overload protection. In a circuit employing a magnetic-only CB, the adjustable magnetic trip element can be set to take over the interrupting task at currents above stalled rotor and up to the short-circuit duty of the supply system at the point of installation. The magnetic trip in a typical unit might be adjustable from 3 to 13 times the breaker current rating. For example, a 100A breaker can be adjusted to trip anywhere between 300 and 1300A. Thus the CB serves as the motor short-circuit protection and motor circuit disconnect.

The following example gives the basic idea of how these type of circuit breakers are to be applied.

GIVEN: A 30-hp, 230V, 80A FLC, 3-phase, squirrel-cage motor marked with a code letter indicating that the motor has a locked-rotor current of 10 to 11.19 kVA per horsepower. A full-voltage, across-the-line controller with overload protection in the controller to protect the motor within its heating damage curve is used.

REQUIRED: Select a circuit breaker that will provide short-circuit protection and will qualify as the motor circuit disconnect means.

THERMAL-MAGNETIC SOLUTION: A CB suitable for use as disconnect must have a current rating at least

$$115\% \times 80 = 92A.$$

The code permits the use of an inverse-time (thermal-magnetic) CB rated not more than 250% of motor full-load current (although a CB could be rated as high as 400% of full-load current if such size were necessary to pass motor starting current without opening). Based on the calculation:

$$2.5 \times 80 = 200A,$$

a 225A-frame size CB with a 200A trip

setting could be selected. The large size of this CB will generally take the starting current of the motor without tripping either the thermal element or the magnetic element in the CB. The starting current of the motor will initially be:

30 hp x 11.19 kVA per hp x 1000 / 220V x 1.73 = 882A.

The instantaneous trip setting of the 200A CB will be about

200 x 10 = 2000A.

Such a CB will provide protection for grounds and shorts without interfering with motor overload protection.

MAGNETIC-ONLY SOLUTION: Consider use of a 100A CB with thermal and adjustable magnetic trips. The instantaneous trip setting at 10 times current rating would be 1000A, which is above the 882A locked-rotor current. But starting current would probably trip the thermal element and open the CB.

This problem can be solved by removing the CB thermal element and leaving only the magnetic element in the CB. Then the conditions of overload can be cleared by the overload devices in the motor starter, right up to stalled rotor. The magnetic trip would be adjusted to open the circuit instantaneously on currents above, say 1300A (882 x 1.5). But because the value of 1300A is greater than 1300% of the motor full-load current (80 x 13 = 1040A), the maximum setting would be 1000A.

Because the use of a magnetic-only CB does not protect against low-level grounds and shorts in the circuit conductors on the line side of the starter overload relays, such application must be made only where the CB and starter are installed as a combination starter in a single enclosure.

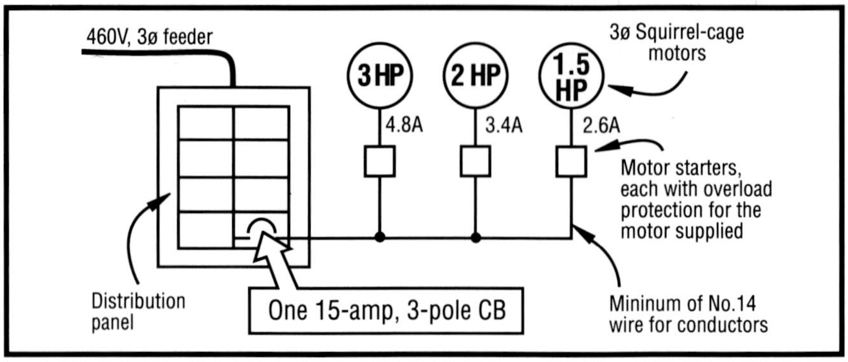

Fig. 5.3. Several motors on one branch circuit, using a circuit breaker for protection.

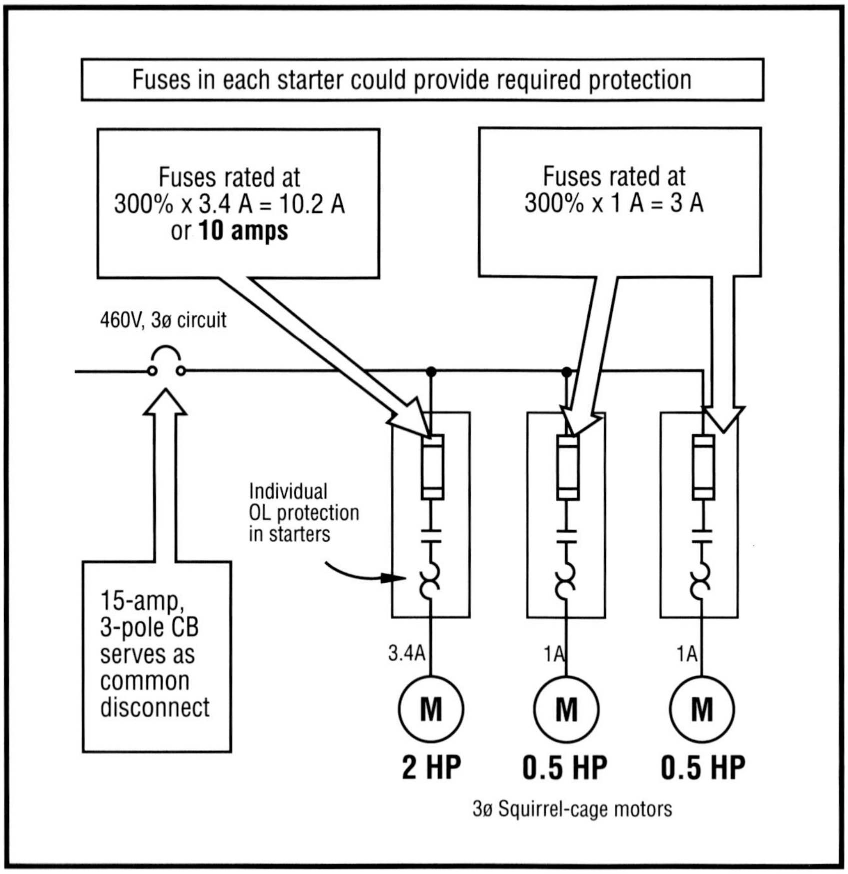

Fig. 5.4. Several motors on one branch circuit, using fuses for protection .

Motor short-circuit protector is a fuse-like device for use only in its own type of combination motor starter. The combination offers short-circuit protection, overload protection, disconnect means, and motor control, all with assured coordination between the short-circuit interrupter (the motor short-circuit protector) and the overload devices. It involves a simple method of selection of the correct unit for a given motor circuit. This packaged assembly is a third type of combination motor starter (in addition to the conventional fused-switch and circuit-breaker types).

The NEC recognizes motor short-circuit protectors provided the combination is identified for the purpose. Practically speaking, this means a combination starter equipped with motor short-circuit protectors and listed by a nationally recognized third-party testing lab.

Multimotor branch circuit. A single branch circuit may be used to supply two or more motors as follows.

Two or more motors, each rated at not more than 1 hp, and each drawing not over 6A full-load current, may be used on a branch circuit protected at not more than 20A at 125V or less, or 15A at 600V or less. The rating of the branch circuit protective device marked on any of the controllers must not be exceeded. Individual overload protection is necessary in such circuits.

Two or more motors of any rating, each having individual overload protection, may be connected to a single

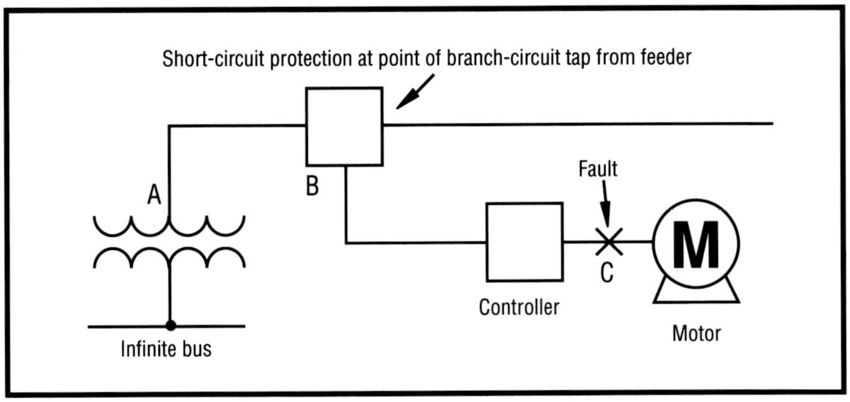

Fig. 5.5. Motor controllers must be protected against "let-through" damage.

branch circuit that is protected by a short-circuit protective device. The protective device must be selected in accordance with the maximum rating or setting that could protect an individual circuit to the motor of the smallest rating. This may be done only where it can be determined that the branch-circuit device so selected will not open under the most severe normal conditions of service that might be encountered.

Fig. 5.3 shows an example of how these rules are applied.

GIVEN: A 3-pole, 15A CB is to be used for the branch-circuit protective device, the FLC for each motor is as shown.

SOLUTION: The rating of the branch-circuit protective device, 15A, does not exceed the maximum value of short-circuit protection required for the smallest motor of the group (the 1 1/2-hp motor). Although 15A is greater than the maximum value of 250% times the motor full-load current (2.5 x 2.6A = 6.5A), the 15A breaker is the next higher size for a standard circuit breaker.

The total load of motor currents is:

4.8 + 3.4 + 2.6 = 10.8A.

This is well within the 15A CB rating, which has sufficient time delay in its operation to permit starting of any one of these motors with the other two already operating. Torque characteristics of the loads on starting are not high. It was therefore determined that the CB will not open under the most severe normal service.

Each motor is provided with individual overload protection in its starter.

Branch-circuit conductors are sized in this case:

4.8 + 3.4 + 2.6 + (.25 x 4.8) = 12A.

Thus, the conductors must have an ampacity at least equal to 12A. No. 14 conductors rated at 75°C will fully satisfy this application.

In **Fig. 5.4**, three motors are hooked-up differently to comply with the rules when fuses, instead of a circuit breaker, are used for branch-circuit protection.

SOLUTION: Fuses used as branch-circuit protection must have a rating not in excess of the value permitted for the smallest motor of the group (one of the 1/2-hp motors).

The maximum permitted rating of nontime-delay type fuses is 300% of full-load current for 3-phase squirrel-cage motors. Applying this rule to one of the 1/2-hp motors (the smallest) gives a maximum fuse rating of

300% x 1A = 3A.

But, there is no permission for the fuses to be rated higher than 3A, because 3A is a standard rating for fuses shown in the code. Thus, the two 1/2-hp motors may be fed from a single branch circuit with three 3A fuses in a 3-pole switch.

Following the same code rules, the 2-hp motor would require:

300% x 3.4A = 10.2A.

Thus, the maximum size standard fuse that can be used is 10A.

An important point to note is that if the maximum branch-circuit short-circuit and ground-fault protective device ratings is marked on the equipment, or provided in literature included with the equipment, these values should not be exceeded even if higher values are allowed by NEC calculations.

In making up motor circuits, a feeder may be tapped by smaller conductors for a motor subfeeder or motor branch circuit without overcurrent protection at the point of tap. NEC rules covers taps up to 10 ft, 25 ft, or 100 ft long.

Let-through current. Branch-circuit protection must always be capable of interrupting the amount of short-circuit current that might flow through it. Also, the speed of clearing the circuit must be compared to the abilities of the various circuit elements to withstand the damaging effects of short-circuit current flow during the time it takes for the protective device to operate.

For example, in **Fig. 5.5**, a short circuit fault at "C" will draw current until the circuit is opened by the protective device at "B." The value of the short-circuit current available at point C depends upon the kVA rating of the supply transformer "A," percent reactance of the transformer, the secondary voltage, and the effective impedance of the current path from the transformer to the point of the fault. Application of motor controllers, therefore, must be coordinated with branch-circuit overcurrent protection that must be able to safely interrupt the short-circuit current. Not only must the device be rated to interrupt the fault current, it must act quickly enough to open the circuit before the let-through current can damage the controller.

Air conditioning and refrigeration. The code covers the rating or setting of the branch-circuit short-circuit and ground-fault protective device for a circuit to an individual sealed hermetic compressor motor. The rule says that the device must be capable of carrying the starting current of the motor. The required protection must be considered as being obtained when this device has a rating or setting not exceeding 175% of the motor-compressor rated-load current or branch-circuit selection current, whichever is greater (15A size minimum). But where the protection specified is not sufficient for the starting current of the motor, it may be increased but must not exceed 225% of the motor rated-load current or branch-circuit selection current, whichever is greater.

The rules in this article also cover

sizing of the short-circuit and ground-fault protective device for a branch circuit to equipment that incorporates more than one sealed hermetic motor-compressor and other motors or other loads.

Article 440 describes a room air conditioner as an AC appliance of the air-cooled window, console, or in-wall type, with or without provisions for heating, installed in the conditioned room, and incorporating one or more hermetic refrigerant motor-compressors. A room air conditioner is treated as a single motor unit in determining its branch-circuit requirements when all the following conditions are met.

- The unit is cord-and-plug connected.
- Its total rating is not more than 40A and 250V, single phase.
- Total rated-load current is shown on the unit nameplate rather than individual motor currents.
- The rating of the branch-circuit, short-circuit, and ground-fault protective device does not exceed the ampacity of the branch-circuit conductors or the rating of the receptacle, whichever is less.

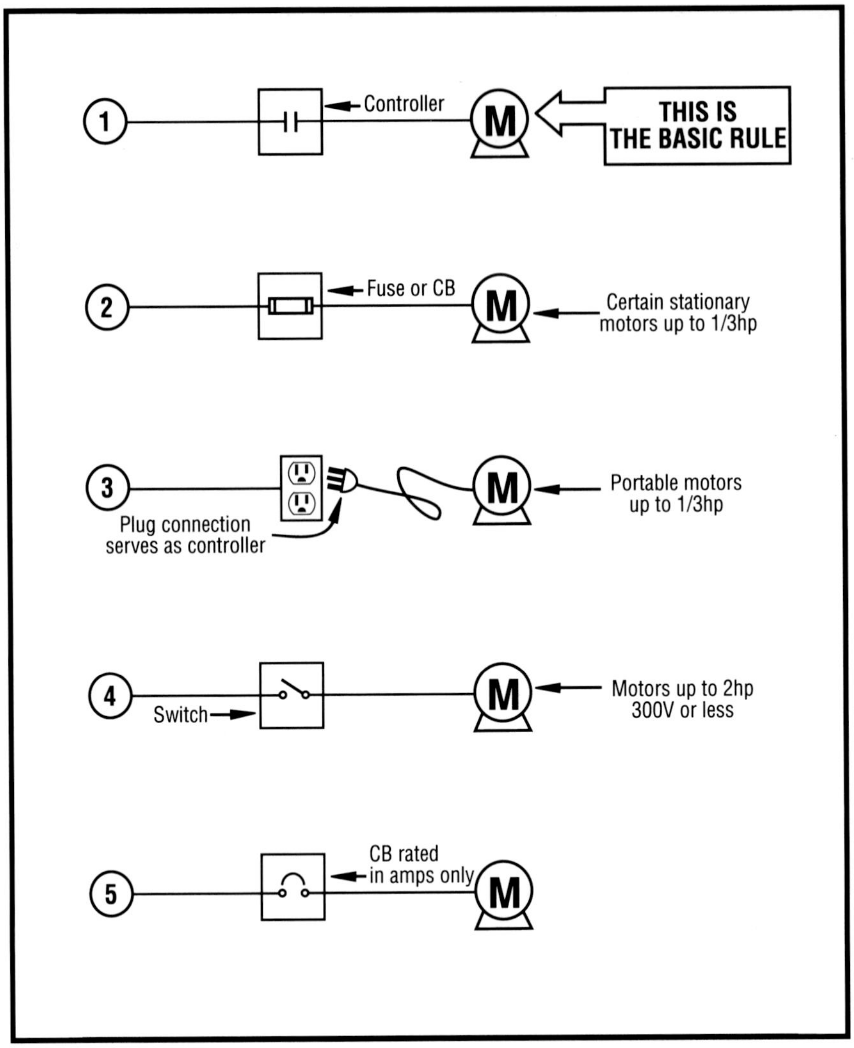

Fig. 5.6. Controller must be a switch, starter, or other device.

3. SELECTING A MOTOR CONTROLLER

According the the definition in the NEC, a contoller is a device or group of devices that serves to govern, in some predetermined manner, the electric power delivered to the apparatus to which it is connected. As used in the code, the term "controller" includes any device normally used to start and stop a motor. A starter consisting of a contactor and overload relay is considered to be a controller. A properly rated snap switch that is permitted to turn a single-phase motor on and off is also considered to be a motor controller. These switches are permitted to serve as the motor controller as well as providing the disconnecting means. Solid-state, adjustable-speed, and DC drives are also considered to be motor controllers.

The contactor of a magnetic starter (not the pilot device) is the controller. In a solid-state starter or an AC or DC drive, it is the power-circuit elements such as SCRs that meet the definition as being the controller. A pushbutton station, a limit switch, a float switch or any other pilot/control device that carries the electric signals directing the performance of the controller is not the controller.

A controller must be capable of starting and stopping the motor that it controls, must be able to interrupt the stalled-rotor current of the motor, and must have a horsepower rating not lower than the rating of the motor. **Fig. 5.6** shows various exceptions to this general rule.

Hp-rated switches are permitted as controllers and motor disconnect means, and horsepower-rated switches up to 500 hp, 600V have been listed. However, instructions for listed horsepower-rated switches given in the UL White Book specifically state that "enclosed switches rated higher than 100 horsepower are restricted to use as motor disconnecting means and are not for use as motor controllers." But a hp-rated switch up to 100 hp may be used as both a controller and disconnect if it breaks all ungrounded legs of conductors to the motor.

HVAC hermetic motors. For sealed (hermetic-type) refrigeration compressor motors, selection of the size of controller is slightly more involved than it is for standard applications. Because of their low-temperature operating conditions, hermetic motors can handle heavier loads than general-purpose motors of equivalent size and rotor-stator construction. And because the capabilities of such motors cannot be accurately defined in terms of horsepower, they are rated in terms of full-

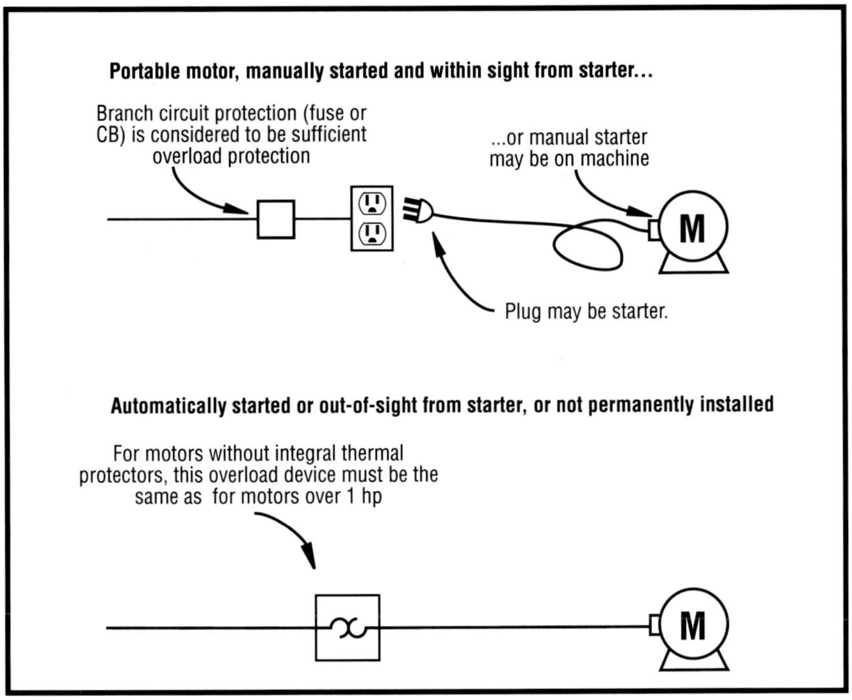

Fig. 5.7. Selection of overload protection for motors rated 1 hp or less.

load current and locked-rotor current for polyphase motors and larger single-phase motors. Accordingly, selection of controller size is different than in the case of a general-purpose motor where hp ratings must be matched.

Code rules on controllers for motor-compressors are covered in Article 440. The controller must have both a continuous-duty full-load current rating and a locked-rotor current rating not less than the full-load and locked-rotor currents of the motor.

For controllers rated in horsepower, selection of the size required for a particular hermetic motor can be made after the nameplate rated-load current, or branch-circuit selection current (whichever is greater) and locked-rotor current of the motor have been converted to an equivalent horsepower rating.

To get an equivalent horsepower rating, which is the required size of controller, the tables in Article 430 must be used. In all tables, if the exact value of current is not listed, the next higher value should be used. If the two horsepower ratings obtained in this way are not the same, the larger value is taken as the required size of controller.

Here is a typical example of the selection process.

GIVEN: A 230V, 3-phase, squirrel-cage induction motor in a compressor has a nameplate full-load current rating of 25.8A and a nameplate locked-rotor current of 90A.

PROCEDURE: From Article 430 FLA table, 28A is the next higher current to the nameplate current of 25.8, and the corresponding horsepower rating for a 230V, 3-phase motor is 10 hp.

From the Article 430 conversion table of locked-rotor currents, a locked-rotor current rating of 90A for a 230V, 3-phase motor requires a controller rated at 5 hp. The two values of horsepower obtained are not the same, so the higher rating is selected as the acceptable unit for the conditions. A 10-hp motor controller must be used.

Some controllers may be rated not in horsepower but in full-load current and locked-rotor current. For use with a hermetic motor, such a controller must have current ratings equal to, or greater than, the nameplate full-load current and locked-rotor current of the motor.

Starter Poles. The NE Code says that a controller need not open all conductors to a motor, except when the controller serves also as the required disconnecting means. For instance, a 2-pole starter could be used for a 3-phase motor if overload protection is provided in all three circuit legs by devices separate from the starter. The controller must interrupt only enough conductors to be able to start and stop the motor.

However, when the controller is a manual (nonmagnetic) starter, or is a manually operated switch or CB (as permitted by the code), the controller itself also may serve as the disconnect means if it opens all ungrounded conductors to the motor. This eliminates the need for another switch or CB to serve as the disconnecting means. But only a manually operated switch or circuit breaker may serve such a dual function. A magnetic starter cannot serve as the disconnecting means, even if it does open all ungrounded conductors to the motor.

Generally, an individual motor controller is required for each motor. However, for motors rated not over 600V, a single controller rated at not less than the sum of the horsepower ratings of all of the motors of the group may be used with a group of motors in any one of the following cases:

- If a number of motors drive several parts of a single machine or piece of apparatus such as metal- and woodworking machines, cranes, etc.
- If two or more motors are under the protection of one overcurrent device as in the case of small motors supplied from a single branch circuit. This use of single controller applies only to cases involving motors of one hp or less.
- If a group of motors is located in one room and all are within sight from the controller location.

4. PROTECTING AGAINST OVERLOADS

The code has specific requirements that apply to motor overload protection intended to protect the elements of the branch circuit, including the motor itself, the motor control apparatus, and the branch-circuit conductors against excessive heating due to motor overloads. When an overload persists for a sufficient length of time, it will cause damage or dangerous overheating of the apparatus. An overload is considered to include stalled-rotor current.

Overload does not include fault current due to shorts or grounds.

Overload devices (relays and thermal cutouts, etc.) are required in each leg of circuits to 3-phase motors unless the motor is protected by other approved means, such as special embedded detectors with or without supplementary external protective devices.

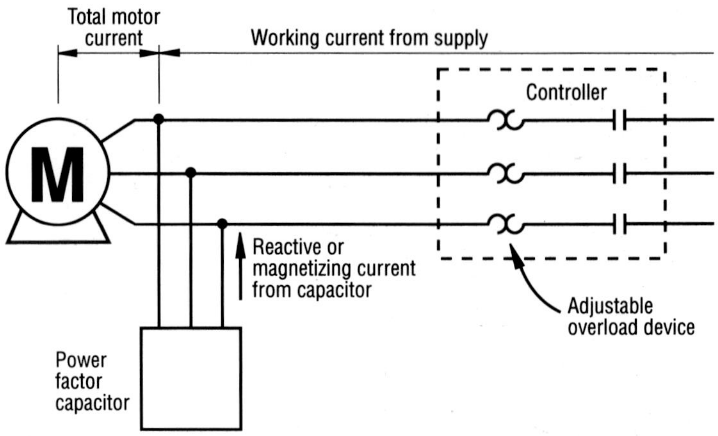

EXAMPLE:
Motor with 70% power factor has full-load current of 143 A. Capacitor corrects to 100% PF, determine setting of OL relays.

SOLUTION:

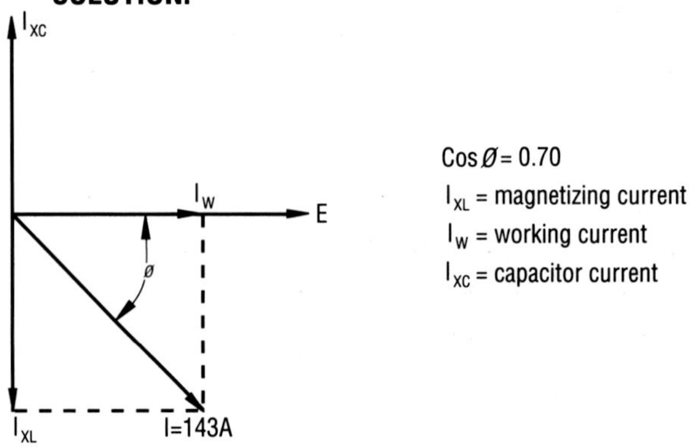

I_{XC} cancels I_{XL} leaving only working current to be supplied from circuit.
Working current = 143 x CosØ = 143 x 0.70 = 100A.

Overload relays should be set for 125% of 100 amps.

Fig. 5.8. Correcting overload setting when capacitors are installed in a motor circuit.

Typical code requirements for overload protection vary according to the size and use of the motor.

Motors more than 1 hp. For a motor that is over 1 hp, if used for continuous duty, overload protection must be provided. This may be an external overcurrent device actuated by the motor running current and set to open at not more than 125% of the motor full-load current for motors marked with a service factor of not less than 1.15 and for motors with a temperature rise not over 40°C.

Sealed (hermetic-type) refrigeration compressor motors must be protected against overload and failure to start by one of the following: an overload relay; an approved, integral thermal protector; a branch-circuit fuse or CB rated at not over 125% of rated load current; a special protective system.

The overload device must be rated or set to trip at not more than 115% of the motor full-load current for all other motors, such as motors with a 1.0 service factor or a 55°C rise.

Motors of 1 hp or less. When motors of less than 1 hp are not permanently installed and are manually started, they are considered protected against overload by the branch-circuit protection if the motor is within sight from the starter. Overload devices are not required in such cases. A distance of over 50 ft is considered out of sight. **Fig. 5.7** refers to these requirements.

Any motor of 1 hp or less that is not portable, is not manually started, and/or is not within sight from its starter location must have specific overload protection. Automatically started motors of 1 hp or less must be protected against overload in the same way as motors rated over 1 hp. That is, a separate or integral overload device must be used.

Exceptions. There are exceptions to the basic rules on providing specific overload protective devices for protection against overloads.

Where the values specified for motor overload protection do not permit the motor to start or to carry the load, the next higher size of overload relay may be used, but not higher than the percentages of motor full-load current rating given in code under "Selection of Overload Relay."

Fuses or circuit breakers may be used for overload protection but may not be rated or set up to those values. Fuses and breakers must have a maximum rating as shown for motors of more than 1 hp. If the value determined as indicated there does not correspond to a standard rating of fuse or CB, the next smaller size must be used.

Under certain conditions, no specific overload protection needs to be used. For example, the motor is considered to be properly protected if it is part of an approved assembly that does not normally subject the motor to overloads and has controls to protect against stalled rotor. Also, if the impedance of the motor windings is sufficient to prevent overheating due to failure to start, the branch-circuit protection is considered adequate.

A motor used for a condition of service that is inherently short-time, intermittent, periodic, or varying duty is considered as protected against overload by the branch-circuit overcurrent

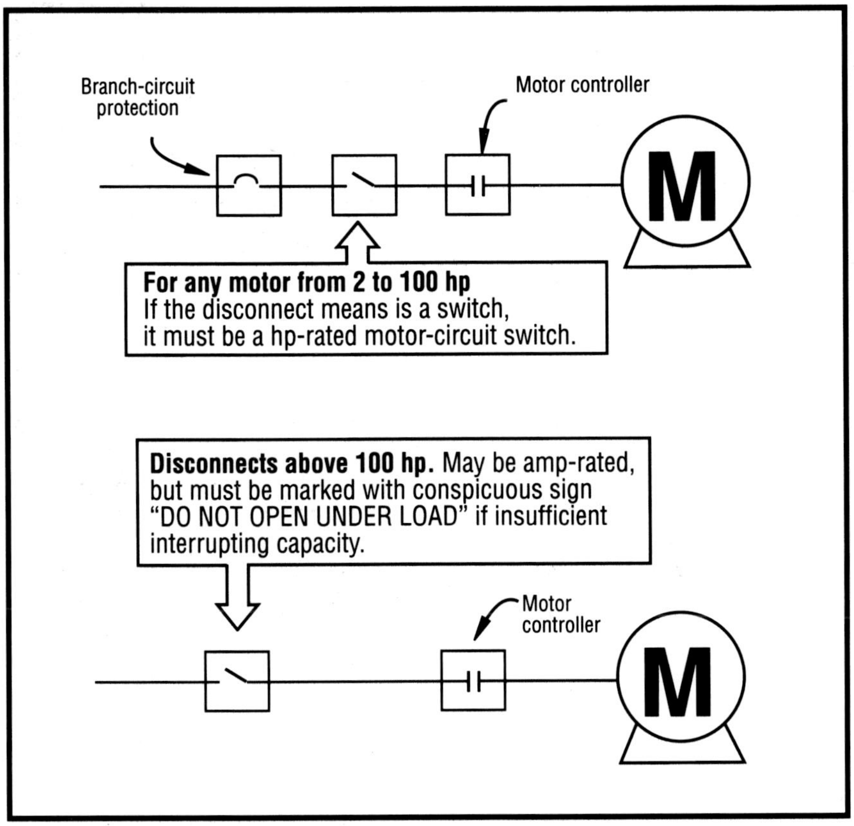

Fig. 5.9. Watch the hp and ampere ratings of disconnects!

device. Motors are considered to be for continuous duty unless the motor cannot operate continuously with load under any condition of use.

Special considerations. Any motor that is automatically started is permitted to have its overload protection shunted or cut-out during the starting period. This accommodates those motor-and-load applications that have a long accelerating time and would otherwise require an overload device with such a long trip-time that the motor would not be protected if it stalled while running. However, conditions are given in the code for shunting-out overload protection of a motor.

Fig. 5.8 outlines the need for correcting the size of overload protection in motor controllers when power-factor capacitors are used on the load side of the controller for correcting power factor.

Although the NE Code has all of those requirements on use of overload protection of motors, it does recognize that there are cases when automatic opening of a motor circuit due to overload may be objectionable from a safety standpoint. In recognition of the needs of many industrial applications, alternatives to automatic opening of a circuit in the event of overload are permitted. When automatic opening of the circuit on an overload would constitute a more serious hazard than the overload itself, the rule notes that automatic overload opening is not required. However, it is necessary that the circuit be provided with a motor overload sensing device conforming to the code requirement on overload protection to indicate by means of a supervised alarm the presence of the overload. Overload indication instead of automatic opening will alert personnel to the objectionable condition and will permit corrective action, either immediately or at some more convenient time, for an orderly shutdown to resolve the difficulty.

5. PROVIDING A DISCONNECT

The code requires that a means (a motor-circuit switch rated in hp, or a CB) must be provided in each motor circuit to disconnect both the motor and its controller from all ungrounded supply conductors. In a motor branch circuit, the disconnect switch or CB must be rated to carry at least 115% of the nameplate current rating of the motor for circuits up to 600V.

A circuit-breaker switching device with no automatic trip operation, is recognized. The "molded-case switch" may be used as a motor disconnect instead of a conventional circuit breaker or a horsepower-rated switch. Such a device either must be rated for the horsepower of the motor it is used with or must have an appropriate ampere rating.

The NE Code makes a basic requirement that the disconnecting means for a motor and its controller be a motor-circuit switch rated in horsepower (see **Fig. 5.9**). This rule can be readily complied with, inasmuch as listed motor-circuit switches up to 500 hp are available, and the manufacturers mark switches to conform. But, for motors rated over 100 hp, the code does not require that the disconnect have a hp rating. It makes an exception to the basic rule and permits the use of ampere-rated switches or isolation switches, provided the switch has a carrying capacity of at least 115% of the nameplate current rating of the motor. And testing lab listing notes say that hp-rated switches over 100 hp must not be used as motor controllers.

Isolation switches for motors over 100 hp, not capable of interrupting stalled-rotor currents, must be plainly marked "DO NOT OPEN UNDER LOAD."

Sizing a disconnect. An example illustrates how a disconnect is to be sized for a motor circuit.

PROBLEM: Provide a disconnect for a 125-hp, 3-phase, 460V motor. Use a non-fusible switch since short-circuit protection is provided at the supply end of the branch circuit. The FLC of the motor is 156A.

SOLUTION: A suitable disconnect must have a continuous carrying capacity of

156 x 1.15 or 179A.

This calls for a 200A, 3-pole switch rated for 480V. The switch could be a general-use switch, a current-and-hp marked motor-circuit switch, or an isolation switch.

A motor-circuit switch with the required current and voltage rating for this case would be marked for 50 hp, but the horsepower rating is of no concern because the switch does not have to be hp-rated for motors larger than 100 hp. If the 50-hp switch were of the heavy-duty type, it would have an interrupting rating of 10 x 65A = 650A (the assumed full-load current of a 460V, 50-hp motor). But the locked-rotor current of the 125-hp motor might run as high as 900A. In such a case, the switch must be marked "DO NOT OPEN UNDER LOAD".

If a fusible switch had been used for the above motor to provide disconnect and short-circuit protection, the size of the switch would be determined by the size and type of fuses used. Using a fuse rating of 250% of motor current for standard fuses, the application would call for 400A fuses in a 400A switch. This switch would certainly qualify as the motor disconnect. However, if time-delay fuses are used, a 200A switch would be large enough to take the time-delay fuses and could be used as the disconnect (because it is rated at least 115% of motor current).

In the foregoing, the 400A switch might have an interrupting rating high enough to handle the locked-rotor current of the motor. Or the 200A switch might be of the type that has an interrupting rating up to 12 times the rated load current of the switch itself. In either of these cases, there would probably be no need for the "DO NOT OPEN UNDER LOAD" marking.

Serving as disconnect and controller. Up to 100 hp, a switch that satisfies the code on rating for use as a motor controller may also provide the required disconnect means—the two functions being performed by the one switch—provided it opens all ungrounded conductors to the motor, is protected by an overcurrent device (which may be the branch-circuit protection or may be fuses in the switch itself) and is a manually operated airbreak switch or an oil switch not rated over 600V or 100A.

A single circuit breaker may also serve as controller and disconnect. However, in the case of an autotransformer type of controller, the controller

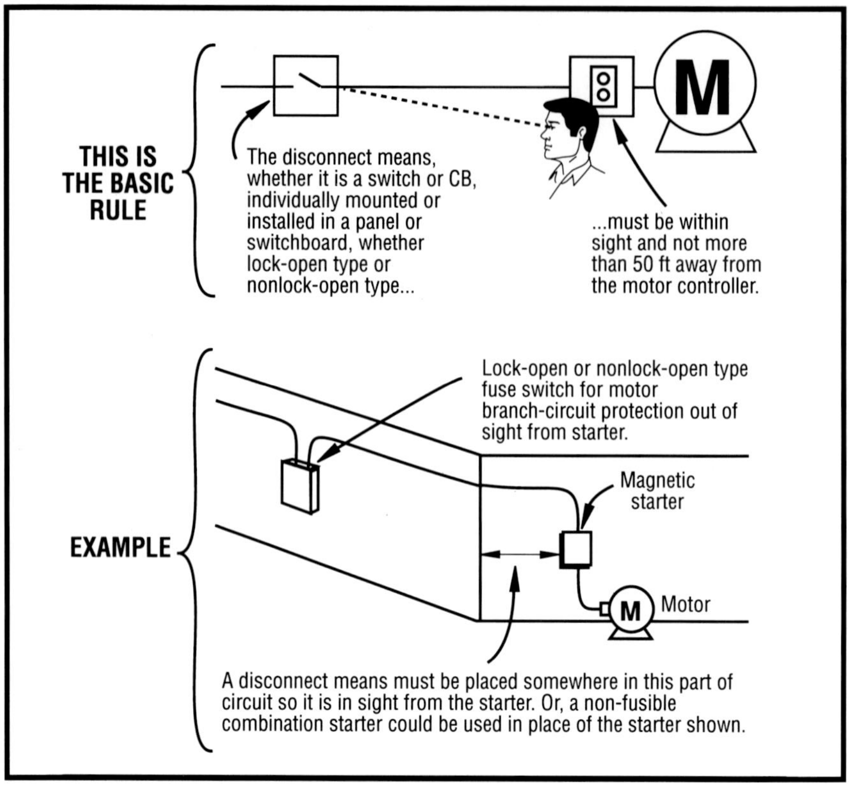

Fig. 5.10. Disconnect locations for motors up to 600V.

itself, even if manual, may not also serve as the disconnecting means. Such controllers must be provided with a seperate means for disconnecting controller and the motor.

The acceptability of a single switch for both the controller and disconnecting means is based on the single switch satisfying the code requirements for a controller and for a disconnect. It finds application where general-use switches are used, as permitted by the code, in conjunction with time-delay fuses rated low enough to provide both overload protection and branch-circuit (short-circuit) protection. In such cases, a single fused switch may serve a total of four functions: as a controller, a disconnect, branch-circuit protection, and as overload protection. It is possible for a single circuit breaker to also serve these four functions.

A single disconnect sometimes may serve a group of motors. Such a disconnect must have a rating sufficient to handle a single load equal to the sum of the horsepower ratings or current ratings. The single disconnect may be used for a group of motors driving different parts of a single piece of apparatus, for several motors on one branch circuit, or for a group of motors in a single room within sight from the disconnect location.

Disconnect location. The NEC specifically requires that a disconnecting means must be provided in each motor circuit in sight from the controller. **Fig. 5.10** shows the basic rule on "in-sight" location of the disconnect means. The NEC further makes it clear that a pushbutton or similar device in the control circuit that energizes and deenergizes the starter coil is not a disconnect device.

In a magnetic motor starter, it is the contactor that actually governs the electric power delivered to the motor to which it is connected, not the pushbuttons. The pushbuttons do not carry the main power current that is delivered to the motor by the contactor. Therefore, it is the contactor that is defined as being the controller.

It is well established that the intent of the code rule, as well as the letter of the rule, is to designate the contactor and not the pushbutton station as the controller, and the disconnect must be within sight from it and not from a pushbutton station or some other re-

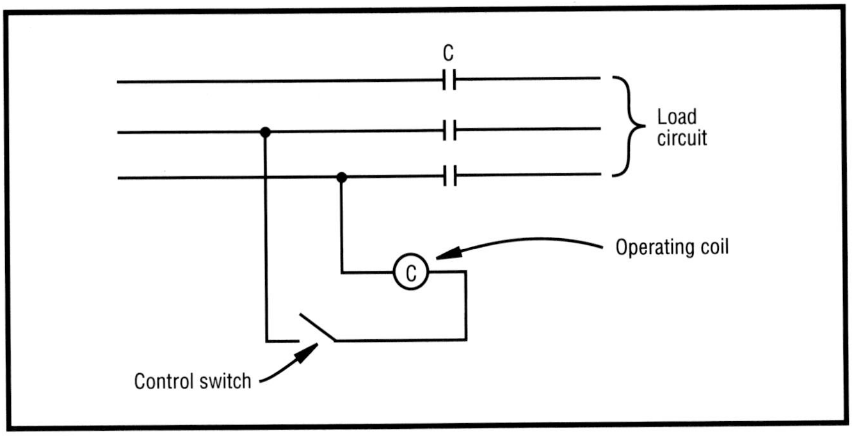

Fig. 5.11. Control circuit used to switch a load circuit.

motely located pilot control device. There are two exceptions to this basic code rule requiring a disconnect switch or CB to be located in sight from the controller.

• The disconnect for a medium-voltage (over 600V) motor is permitted to be out of sight from the controller location.

• Disconnect means for industrial applications of large and complex machinery utilizing a number of motors to power the various interrelated parts of the machine are permitted to be out-of-sight.

An exception to the general rule recognizes that a single common disconnect for a number of controllers is often impossible to be installed within sight of all the controllers even though the controllers are adjacent one to the other. On much industrial process equipment, the components of the overall structure obstruct the view of many controllers. The exception permits the single disconnect to be technically out of sight from some or even all the controllers if the disconnect is simply adjacent to them or nearby on the equipment structure.

An additional rule requires a disconnect means to be within sight and not more than 50 ft away from the motor location and the driven machinery location. But the Exception to that basic requirement says that a disconnect does not have to be within sight from the motor and its load if the required disconnect ahead of the motor controller is capable of being locked in the open position. The intent here is to permit maintenance workers to lock the disconnecting means ahead of the controller in the open position and keep the key in their possession so that the circuit cannot be energized while they are working on it.

Note that the code provisions are minimum safety requirements. Additional use of disconnects, with and without lock-open means, may be made necessary or desirable by job conditions.

6. PROVIDING A CONTROL CIRCUIT

Code rules on motor-starter control circuits must be evaluated against a number of very important background facts.

Definition. A control circuit is any circuit that has as its load device the operating coil of a magnetic motor starter, a magnetic contactor, or a relay. It is a circuit that exerices control over one or more other circuits. These other circuits controlled by the control circuit may themselves be control circuits or they may be "load" circuits, carrying utilization current to a lighting, heating, power, or signal device. **Fig. 5.11** clarifies the distinction between control circuits and load circuits.

Applicable rules. The NEC has no specific rule that covers ampacity of control-circuit wires with respect to the amount of current a starter operating coil draws. The code simply presumes that the wires of a control circuit will have ampacity sufficient for the coil current.

Code rules require protection of control-circuit wires against current in excess of their ampacity or spell out very specific conditions under which overcurrent devices rated higher than wire ampacity may be used, or under which protection of control wires may be, or must be, eliminated.

NEC rules on control-circuit overcurrent protection are not concerned with protecting the operating coil itself against overcurrent. Because of that, it is permissible to select the size of control-circuit wires solely in relation to the rating of branch-circuit protection when use of larger control wire would eliminate the need for a control-circuit fuseblock in the starter.

When a control circuit derives its current supply from the same branch that supplies power to the motor (either directly by being tapped from the line terminals within the starter, or indirectly by being fed from the secondary of a control transformer that has its primary tapped from the starter line terminals and located within the starter enclosure), all of the applicable NEC Article 430 rules must be observed. But where coil-circuit current is derived from a panelboard or from a control transformer that is not part of the starter assembly and is not fed by the motor power circuit, the applicable rules are found in Article 725.

The elements of a control circuit include all of the equipment and devices concerned with the function of the circuit: conductors, raceway, contactor operating coil, source of energy supply to the circuit, overcurrent protective devices, and all switching devices that govern energization of the operating coil.

Design and installation of control circuits are basically divided into three classes in Article 725 according to the energy available in the circuit. Class 2 and 3 control circuits have low energy-handling capabilities.

The vast majority of control circuits for magnetic starters and contactors could not qualify as Class 2 or Class 3 circuits because of the relatively high energy required for operating coils. Any control circuit rated over 150V can never qualify, regardless of energy.

Class 1 control circuits include all

operating coil circuits for magnetic starters that do not meet the requirements for Class 2 or Class 3 circuits.

Control wires in raceway. Class 1 control wires may be run in raceways by themselves. A given conduit, for instance, may carry one or several sets of Class 1 control wires. And use of Class 1 wires must conform to the same basic rules from NEC Chapter 3 that apply to standard power and light wiring.

Conduit fill must be made in accordance with Article rules that cite the usual rules on number of wires in rigid metal conduit, in IMC, in EMT, etc. When more than three control wires are installed in a raceway, load-current-limiting factors need to be applied only if the conductors carry continuous loads (for 3 hrs or more) that exceed 10% of the ampacity of each conductor.

The basic rule on coil circuits says that wires larger than No. 14 must be protected at their ampacities without derating factors if the continuous current on the coil-circuit wires is not in excess of 10% of the ampacity of each conductor.

Two specific sections of the NEC cover the use of Class 1 control conductors in the same raceway, cable, or enclosure containing the circuit wires carrying power to the motor windings. **Fig. 5.12** shows three cases where Class 1 control wires are run in raceway with the motor power conductors. Although the sketches show specific hookups of equipment, the rules described apply to any hookup or equipment layout.

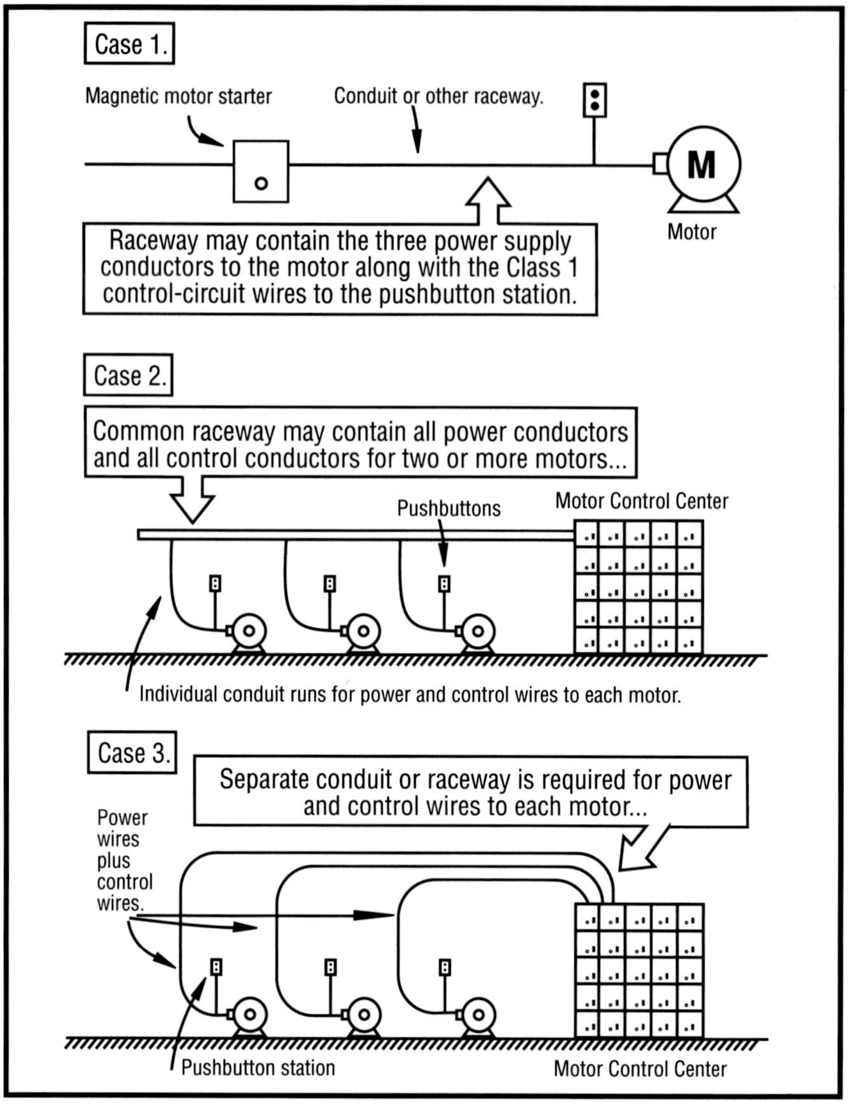

Fig. 5.12. Running starter coil-circuit wires in raceways. Case 1 is permitted by Sec. 725-15; in Case 2, Sec. 300-3(a) permits common raceway to contain; power and control wires of two or more motors if functionally associated; but in Case 3, Sec. 725-15 prohibits intermixing power and Class 1 wiring when motors are not functionally associated.

CASE 1. Running the power and control wires in the same raceway for an individual motor is acceptable because the control circuit is "functionally associated" with the power circuit equipment.

CASE 2. NEC wording recognizes the use of power and control wires in a single raceway to supply more than one motor, but such usage must be made to conform with specific rules.

A common raceway may be used only where the two or more motors are required to be operated together to serve their load function. In such cases, either all motors operate or none do. Running control wires and power wires in the same raceway does not produce a situation where a fault in one motor circuit could disable another circuit to a motor that must otherwise be kept operating.

CASE 3. For those cases where each motor is serving a separate, independent load, the use of a separate raceway for each motor is required —but only when control wires are carried in the raceways. For the three motors shown, it would be acceptable to run the power conductors for all of the motors in a single raceway and all of the control circuit wires in another raceway. Such hookup would not violate the rules, although deratings would have to be made if the continuous current on any control conductor exceeds 10% of the ampacity of that conductor and there is the definite chance of loss of more than one motor on a fault in any one of the circuits in either the power raceway or the control raceway.

Note that an exception in Article 725 permits power and control wires for more than one motor in a common raceway. Power and Class 1 control conductors that are not functionally associated may be grouped in certain cases. This exception recognizes the use of listed motor control centers that have power and control wiring in the same wireway or gutter spaces. The

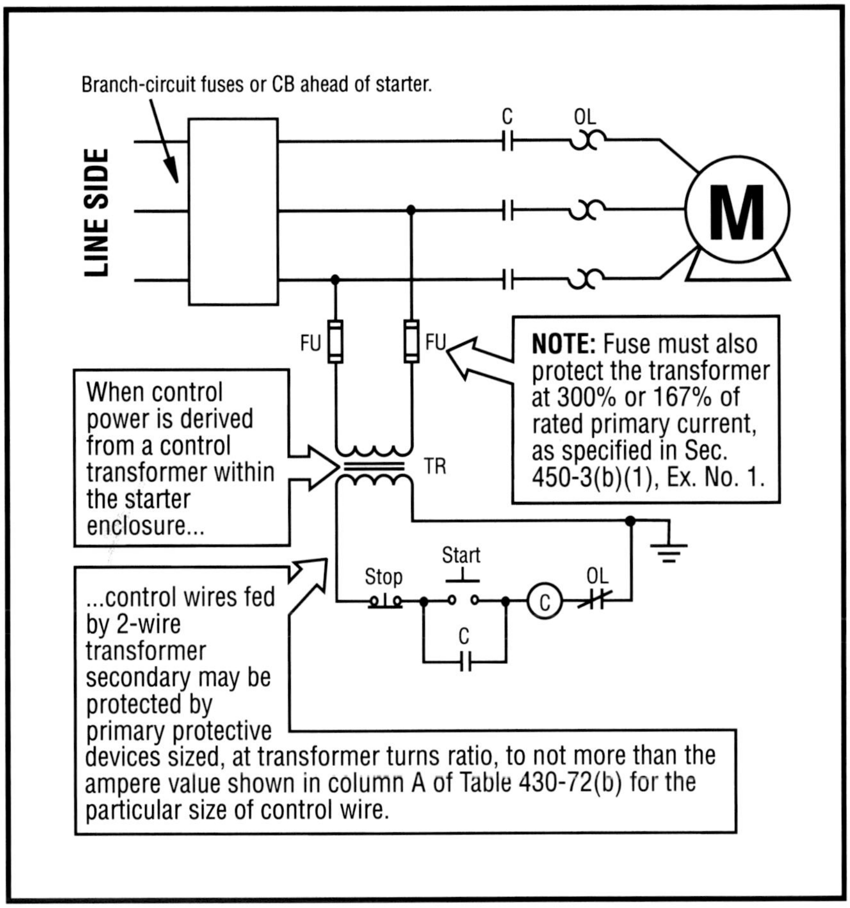

Fig. 5.13. Primary protection may be used for secondary control wires from a control transformer supplying coil current.

basic rule generally prohibits that condition when hooking up motor circuits.

7. PROTECTING CONTROL CIRCUITS

In magnetic motor controllers, the voltage of the operating coil is derived from two of the conductors supplying the load. The line voltage can be used in the control circuit, or in most cases it is desirable to use control circuits and devices of lower voltage rating than the motor. In such cases, control-circuit transformers are used to step the voltage down. There are many NEC rules that cover control-circuit overcurrent protection.

Overcurrent protection. In general, remote-control conductors must be protected against overcurrent. Article 430 covers overcurrent protection for remote-control conductors that are tapped from the line terminals within a motor starter. Article 725 covers overcurrent protection for remote-control conductors of motor starter coil circuits that are derived from separate control transformers or from panelboards, and overcurrent protection for the coil circuits of magnetic contactors used for control of light or power.

An exception in Article 725 states that No. 14 and larger remote-control conductors, other than those for motor-control circuits, can be satisfactorily protected by overcurrent devices rated at not more than 300% of the carrying capacity of the control-circuit conductors. That applies to control wires for magnetic contactors used for control of lighting or heating loads, but not motor loads.

Article 430 covers a similar requirement for motor control circuits. For motor starter coil circuits tapped from line terminals within the starter enclosure, a number of points must be observed.

- Depending upon conditions, the conductors of the control circuit will be considered as protected either by the branch-circuit protective device ahead of the starter or by supplementary protection (usually fuses) installed in the starter enclosure.
- The control circuit that is tapped from the line terminals within a starter is not a branch circuit itself.
- Overcurrent protection is required for the control conductors and not for the operating coil. Because of this, the size of control conductors can be selected to allow application without separate overcurrent protection.

The basic rule requires coil-circuit conductors to have overcurrent protection rated in accordance with the maximum values given in a table in Article 430. Exceptions to the basic rule cover conditions under which other ratings of protection may be used.

EXCEPTION 1 covers protection of control wires for those magnetic starters that have their START-STOP buttons in the cover of the starter enclosure. The branch-circuit protective device may protect control-circuit conductors if the control conductors do not actually leave the enclosure of a magnetic motor starter.

EXCEPTION 2 covers protection of control wires that run from a starter to a remote-control device (pushbutton station, float switch, limit switch, etc.). Such control wires may be protected by the branch-circuit protective device (without need for separate protection) if the branch-circuit device has a rating not over the value shown for the particular size of copper or aluminum control wire in the table.

Control-circuit transformers. Control transformers are relatively small, compact, dry-type potential transformers. They are available in ratings to meet any common motor-control circuit application. For a very wide range of motor controllers, control transformers are available as accessory equipment to the basic starter types. They can be supplied as separate units or incorporated in the controller enclosure. They can be obtained with fused or otherwise protected secondaries to meet code requirements on control-circuit overcurrent protection. Extra transformer capacity can be included to permit operation of a programmable controller of other significant control-circuit loads.

For low-voltage motor controllers, typical control transformers have single or double primary and secondary windings to give either a basic transformation (480V to 120V), or a selection of transformations (480/240V primary to 240/120V secondary). These units range in capacity from 25VA to as high as 8000VA. Control transformers for medium-voltage controllers for 2300V and 4000V motors are generally built into controller enclosures.

Protecting transformer circuits. When control power for a motor-starter coil circuit is derived from a control transformer within the starter enclosure, Article 430 permits protection on the primary side to protect both the transformer and the secondary conductors. This use is limited to transformers with 2-wire secondaries as shown in **Fig. 5.13.**

When using transformer primary-side protection, the rating of overcurrent protective devices must not be greater than the secondary conductor ampacity times the secondary-to-primary voltage ratio.

An exception eliminates any need for control-circuit protection where opening of the circuit would be objectionable, as for a fire-pump motor or other essential safety-related operations.

Another exception eliminates any need for protection of any control transformer rated less than 50VA, provided it is part of the starter and within its enclosure.

The use of secondary protection for a control transformer is suited to applications where the transformer is located within the starter enclosure as shown in **Fig. 5.14.**

Separate voltage source. Article 725 applies to protection of remote-control conductors fed by the 2-wire secondary of a separate control transformer supplying the coils of one or more motor starters of magnetic contactors. The rule is the same as the one described previously.

A properly-sized circuit breaker or fuses may be used at the supply to the transformer primary to provide

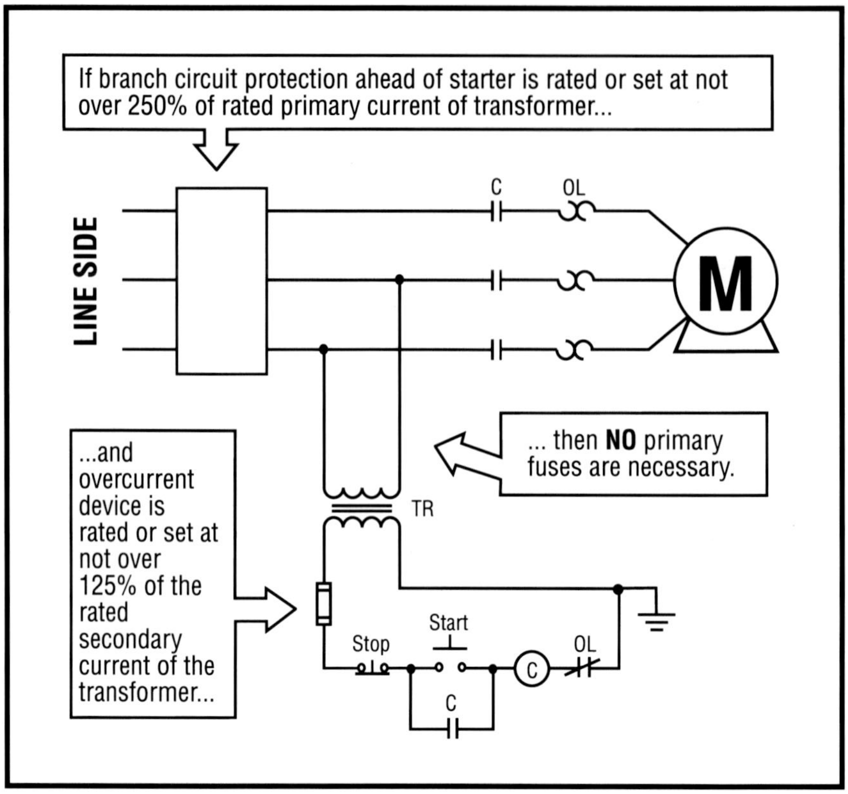

Fig. 5.14. Protection when transformer is in starter enclosure.

overcurrent protection for the primary conductors, for the transformer itself, and for the conductors of the control circuit connected to the secondary.

Primary protection must not exceed the amp rating of the primary circuit conductors. When protection is sized for the transformer, as described previously, No. 14 copper primary conductors will be protected well within their 15A rating.

Secondary conductors for the control circuit can then be selected to have an ampacity at least equal to the rating of primary protection times the primary-to-secondary transformer voltage ratio. Of course, larger conductors may be used if needed to keep voltage drop within limits.

8. OTHER CONTROL-CIRCUIT CONSIDERATIONS

Control circuits associated with motor controls can be extremely complex and vary greatly with the application. Besides the items covered in previously, there are other code requirements that must be adhered to.

Grounding. When a control transformer is used to derive control power, the transformer secondary must be operated with one conductor grounded. Article 250 of the NEC applies to the secondaries of control transformers. According to the rules of this section:

- any 120V, 2-wire circuit must normally have one of its secondary-side conductors grounded;
- the neutral conductor of any 240/120V, 3-wire, single-phase circuit must be grounded; and
- the neutral of a 208/120V, 3-phase, 4-wire circuit must be grounded.

An exception permits ungrounded control circuits under certain specified conditions. A 120V control circuit may be operated ungrounded when all of the following conditions exist.

- The circuit is derived from a transformer that has a primary rating of less than 1000V.
- Supervision will assure that only persons qualified in electrical work will maintain and service the control circuits.
- There is a need for preventing circuit opening on ground fault, such as for safety or for operating reliability.

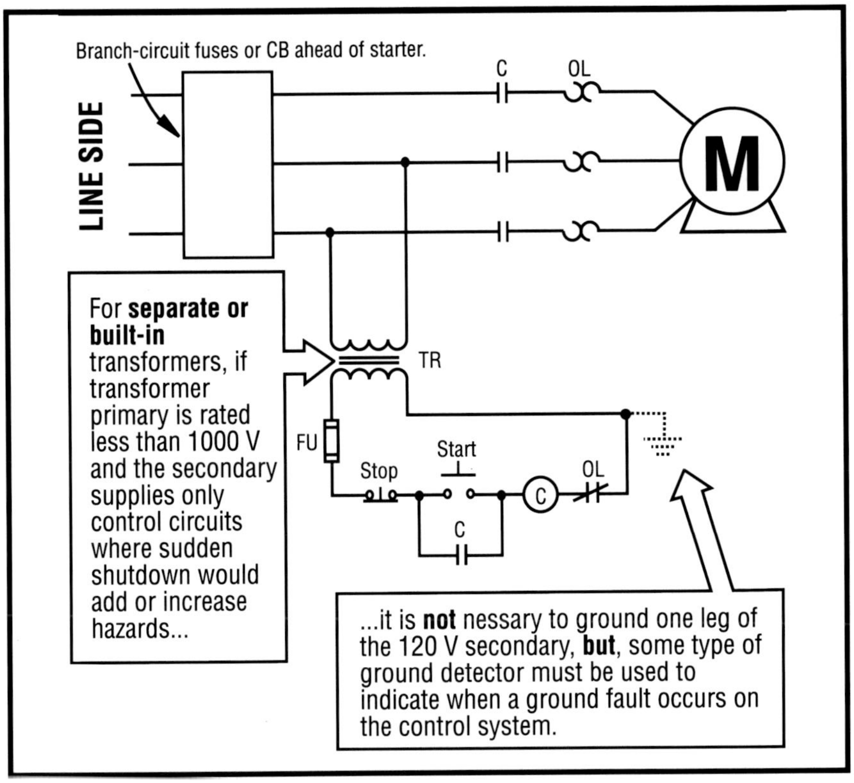

Fig. 5.15. Code rules on grounding and protection apply to control-circuit transformers located either within the starter enclosure or remote from it.

• Some type of ground detector is used on the ungrounded system to alert personnel to the presence of any ground fault, enabling them to clear the ground fault during normal system maintenance.

The NEC rules relating to the grounding of motor control circuit transformers is illustrated in **Fig. 5.15.**

The code rule permitting ungrounded control circuits is primarily significant only for 120V control circuits. The NEC has long permitted 240V and 480V control circuits to be operated ungrounded.

According to the basic rule, when a control transformer has one of its secondary conductors grounded, it is necessary to bond the grounded conductor to the metal case of the transformer and run a grounding electrode conductor from the grounded transformer secondary terminal to nearby grounded building steel of a grounded metal water pipe. But an exception exempts small control transformers from the basic requirement for a grounding electrode conductor run.

A Class 1 remote-control transformer that is rated at not over 1000VA simply has to have a grounded secondary conductor bonded to the metal case of the transformer using a bonding conductor not smaller than No. 14 copper or No. 12 aluminum. No grounding electrode conductor is needed provided that the metal transformer case itself is properly grounded by grounded metal raceway that supplies its primary by means of a suitable equipment grounding conductor that ties the case back to the grounding electrode for the primary system. The connection between the neutral and frame is depended upon to provide an effective return for clearing faults.

Control disconnects. Rules on disconnects for control circuits of magnetic starters require that motor control circuits be so arranged that they will be disconnected from all sources of supply when the disconnecting means for the motor controller is in the open position. However, the rule permits the disconnecting means to consist of two or more separate devices, one of which disconnects the motor controller, and the other the control circuit from its sources of supply. Where separate disconnecting devices are used they must be located immediately adjacent to each other.

In recognition of the unusual and complex nature of large control circuits, Article 430 modifies the basic rule that the disconnecting means must be located immediately adjacent one to the other. When a piece of motor-control equipment has more than 12 motor-control conductors associated with it, remote location the disconnect means is permitted. This permission is applicable only where qualified persons have access to the live parts. Warning signs must be placed on the equipment that locate and identify the various disconnects associated with the control-circuit conductors.

An exception gives another instance in which control disconnects may be mounted other than immediately adjacent to each other. It notes that where the opening of one or more motor control-circuit disconnects might result in a hazard to personnel or property, remote mounting may be used, provided that:

• access is limited to qualified persons; and

• a warning sign is located on the outside of the equipment to indicate the location and identification of each remote control-circuit disconnect.

A similar recognition of the need for remote disconnects in complex layouts is covered. The basic rule of this section requires that a disconnecting means be provided from each source of electrical energy input to equipment having more than one circuit supplying power to it. Each source is permitted to have a separate disconnecting means. But an exception to the code rule states that where a motor receives electrical energy from more than one source (such as a synchronous motor receiving both AC and DC), the disconnecting means for the main power supply to the motor is not required to be immediately adjacent to the motor, provided that the motor controller disconnect is capable of being locked in the open position.

Control circuit arrangement. The design of a control circuit must prevent the motor from being started due to a ground fault in the control circuit wiring. The rule must be observed for any control circuit that has one leg grounded.

By switching the hot leg, the starting of the motor due to an accidental ground fault can be effectively eliminated (**Fig. 5.16**).

This requirement is largely taken care of during the manufacture of standard motor-starting equipment. The wiring of the control circuit components within the starter places the starter coil in the grounded leg. In a typical full voltage, across-the-line, 3-phase starter, all the internal control components are wired to terminal points or blocks. The only accessible points available for external wiring are marked 1, 2, and 3. Terminal 1 is the hot leg; Terminal 2 is the seal-in contact; and Terminal 3 is the coil side of the seal-in contact. Thus, the control wiring is restricted to the hot leg.

Due to a longtime standard convention, the overload relay contact in a standard across-the-line starter is wired on the grounded side of the starter coil. Strictly speaking, this is a violation of the code rule. In actuality, the conductors that tie the starter coil to the overload relay contact are very short, and in some models are integrally bussed arrangements. The likelihood of a ground fault developing in this connection are very remote, as the safety record of this type of equipment has proved.

Another requirement for control-circuit arrangements is that a ground fault will not bypass manual shutdown devices or automatic safety shutdown devices.

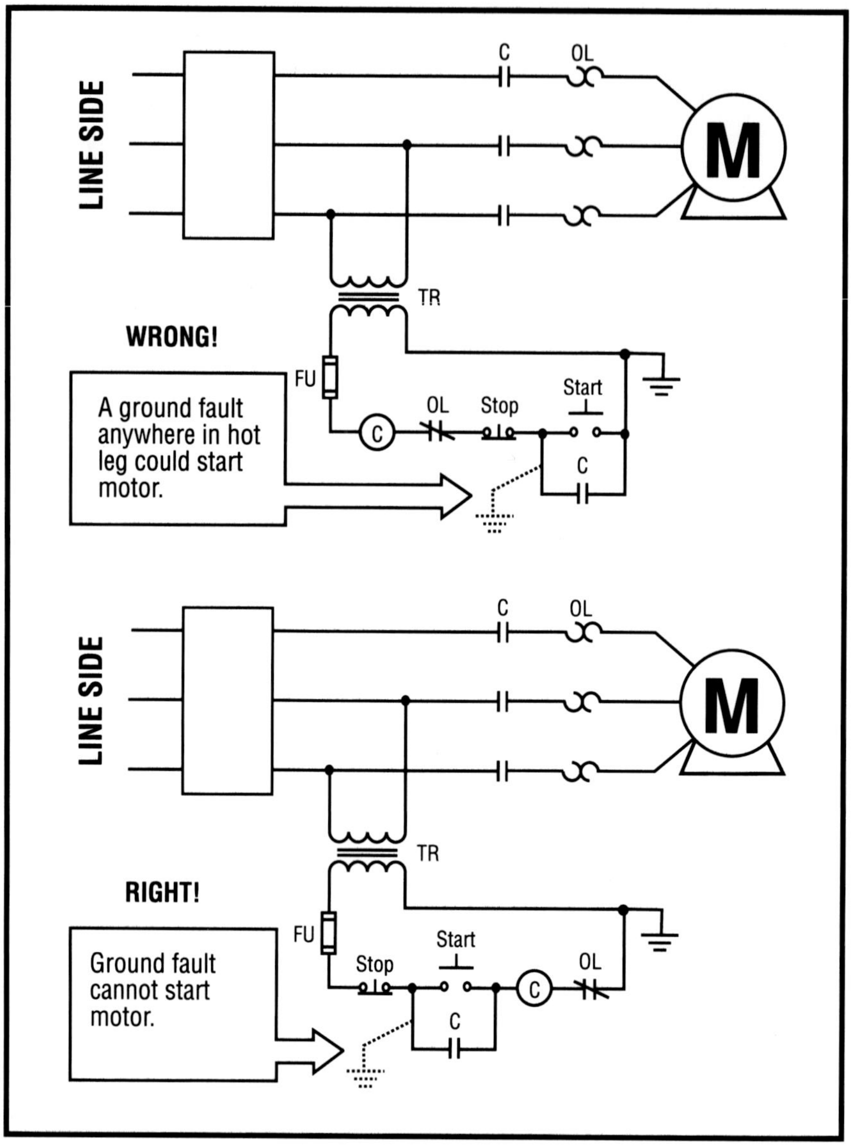

Fig. 5.16. Code rule for the arrangement of a control circuit that has one leg grounded.

SPECIFYING

6

ELECTRIC MOTORS are available in a large number of types, configurations, and constructions. There are DC motors, AC induction motors, wound-rotor motors, 2-speed motors, various types of single-phase motors, and special-application motors such as stepping and linear motors. Each motor has specific associated starting, control, and protective devices that must be specified.

Motors and their controllers or drives rarely operate independently. They are usually part of a larger control scheme that includes sensors, logic devices, monitoring and alarm equipment, operator interface devices, control panels, and many other items. These must all be specified when designing a new or updated facility.

MOTOR TYPES

DC motors are often specified where variable speed is required. However, development of solid-state adjustable-speed drives to control AC motors has provided an alternative for achieving continuous stepless speed variation. Mechanical and magnetic speed changers also can be used for the purpose. Where the application is for single-motor or limited-size systems requiring moderate speed accuracy, a DC drive is adequate. When the application calls for high precision or involves motors in hazardous locations, the adjustable-speed alternative should

Electric motor specifications have to be carefully written to match the motor to its intended load and the environment in which it is to be applied.

be considered.

Polyphase squirrel-cage induction motors constitute the bulk of the motors specified. They are simple, efficient, durable and have an extremely wide range of applications. Use of adjustable-speed drives normally involves a squirrel-cage induction motor; but synchronous-reluctance, and permanent-magnet-rotor synchronous motors can be specified instead when tighter tracking is required.

Wound-rotor motors are used where step-type speed variation, controlled torque, and heavy-duty service are necessary. Cranes and hoists are typical applications.

Synchronous motors are highly efficient units that operate in synchronism with the line frequency. Since they can operate at a leading power factor, they also serve to correct system power factor. Even though they are available in all sizes, their greatest application is in the higher-hp sizes.

Pancake motors are used where the length of an ordinary motor would be unacceptable; machine-tool applications are typical.

MOTOR CHARACTERISTICS

Horsepower. Power requirements are dictated principally by the nature of the load; constant loads have to be treated differently from cyclical loading or intermittent high overloads. It is standard practice to match the motor hp as closely as possible with that of the driven load, since power factor and efficiency increase as the motor approaches its full-load rating. However, exact matching is not likely, since motors are available in standard sizes only.

A way to approach sizing the motor to allow for future load growth is by specifying the motor with a "service factor." Most open induction motors rated 200 hp and lower are supplied with a built-in service factor of 1.15, which means that the motor can deliver 15% more horsepower than its nominal rating. When operating at this overload, the motor will have a higher temperature rise and different efficiency, power factor, and speed; but locked-rotor torque and current and breakdown torque will remain unchanged. Dripproof motors 250 hp and above and totally-enclosed motors normally can be ordered with service factors at higher cost. Accommodating a service factor is usually accomplished by the use of a higher class of insulation to allow for the higher temperature rise.

Voltage and frequency. Standard voltages for 60-Hz polyphase motors are 200, 230, 460, 575, 2300, 4000, 6600, and 13,200V. Specifying a motor at 440 or 480V, unless required for a specific purpose, will result in a nonstandard motor that will probably cost more.

Motors are rated to operate satisfactorily with voltage variations up to 10%, a frequency variation of 5%, and a combined voltage and frequency variation of 10% (provided the frequency variation does not exceed 5%). Operation at voltages higher or lower than these limits can severely reduce insulation life.

The use of a 230V motor on a 208V system is not recommended. In cases where the proper voltage level cannot be supplied, it is better to specify a triple-rated motor (460/230/200V), sometimes available as a standard, or specify a 200V motor.

Speed. Specifying the correct motor speed can avoid having to use a speed reducer or increaser. It is also important to keep in mind the speeds that are standard for the different types of motors.

- Polyphase induction motors are available from 514 to 3600 rpm in smaller sizes and from 225 to 3600 rpm in larger sizes. Nominal speed ratings are 900, 1200, 1800, and 3600 rpm (at 60 Hz).
- Synchronous motors are available in 30 different speed ratings, from 80 to 3600 rpm (at 60 Hz).
- Multispeed motor of a single-winding type with a 2:1 speed ratio.
- Two-winding motor with two independent speeds.
- Pole amplitude motor (PAM), a special 2-speed, single-winding motor with a flexible ratio of low to high speed.

Multispeed motors provide the possibility of achieving staged changes in speed without having to rely on complex DC or variable-speed drives. The technique is often used when speed changes are required during different stages of a process. This approach is often used with mixers. Multispeed motors can be specified in various forms.

- Variable torque, for fans and centrifugal pumps.
- Constant torque, for conveyors, compressors.
- Constant hp for winches and machine tools.

Efficiency. An "energy-efficient" AC motor operates with minimum losses. Its premium cost is returned in six months to three years depending on the operating circumstances. Energy-efficient AC motors should be considered for any application requiring around-the-clock operation. Because of their design, they operate cooler, have longer life, and generally are more reliable.

NEMA DESIGNATIONS

A majority of the motors that are used in the U.S. conform to the specifications of the National Electrical Manufacturers Association (NEMA). Because of this wide compliance, motors of different manufacturers are interchangeable in dimensions; such as from base to centerline of shaft, and others.

There are several definitions that must first be recalled before the specifications can be written.

- Breakdown torque is the maximum torque that will develop at rated voltage and frequency without an abrupt drop in speed.
- Locked-rotor torque is the minimum torque developed when rated voltage and frequency are applied with the rotor at rest and at any angular position.
- Locked-rotor current is the steady-state current drawn by the motor at the moment that rated voltage and frequency are applied to a motor whose rotor is at rest.

Design letters can be used in specifications to indicate desired characteristics of a motor.

Code Letter	Kilovolt-Amperes per Horsepower with Locked Rotor
A	0 – 3.14
B	3.15 – 3.54
C	3.55 – 3.99
D	4.0 – 4.49
E	4.5 – 4.99
F	5.0 – 5.59
G	5.6 – 6.29
H	6.3 – 7.09
J	7.1 – 7.99
K	8.0 – 8.99
L	9.0 – 9.99
M	10.0 – 11.19
N	11.2 – 12.49
P	12.5 – 13.99
R	14.0 – 15.99
S	16.0 – 17.99
T	18.0 – 19.99
U	20.0 – 22.39
V	22.4 – and up

Table 6.1. Locked-rotor indicating code letters [derived from NEC Table 430-7(b)].

• NEMA Design B is the most commonly used type, applicable to a wide variety of equipment loads. Unless specified otherwise, Design B motors will be supplied by manufacturers when filling an order. Slip for Design B motors is limited to 5% or less.

• NEMA Design A is similar to Design B except that they have higher breakdown torques and locked-rotor current.

• NEMA Design C motors have a high starting (locked-rotor) torque and low starting current.

• NEMA Design D motors have high starting torque and slip in the range of 5 to 8% and from 8 to 13%.

Code letter. The NEC requires that the motor nameplate contain a code letter indicating the locked-rotor (starting) kVA per hp. **Table 6.1** shows the code letters as listed in NEC. Code letters are important, since they determine the maximum setting of motor branch-circuit devices. Although this information is usually supplied by the motor vendor to the purchaser, there may be times when the system being designed requires limitations on motor-starting current. In that event, this requirement can be detailed in the specifications.

Insulation. There is a specific temperature rise permitted by standards based upon the insulating material capabilities. A rule-of-thumb is that for every 10°C rise above the limit, insulation life is halved. The total allowable temperature for the different insulation classes (including ambient temperature and temperature rise) are:

• Class A, 105°C,
• Class B, 130°C,
• Class F, 155°C, and
• Class H, 180°C.

Specifiers normally do not have to indicate the type of insulation required; Class B insulation is considered standard and will most often be supplied. A specification for a 1.15 service factor for a totally-enclosed motor will usually be met by supplying a higher grade of insulation. There are cases, however, when a higher insulation class is justified as a safety factor. Operating an open-dripproof motor in an area with high ambient temperature can be compensated for by reducing the service factor or supplying a higher-rated insulation.

Permitted temperature rise of different insulations is based on the operation of the motor at altitudes of 3300 ft (1000 m) or less. When this elevation must be exceeded, there are several alternatives. If the motor has a 1.15 service factor, then the motor can be operated at unity service factor at altitudes up to 9000 ft in a 40°C ambient.

Reduction of the ambient temperature also acts to compensate for increases in altitude. For instance, for Class B insulation, the permissible temperature rise (80°C at unity service factor for altitudes up to 3300 ft) is maintained up to 6600 ft if the ambient temperature is only 30°C. Similarly, a 20°C ambient temperature permits the same temperature rise up to an altitude of 9900 ft.

ENCLOSURES

Environmental conditions under which a motor operates is the determining factor in specifying the proper motor enclosure.

Dripproof motors Dripproof motors have ventilating openings arranged to prevent water or solid particles (falling at not greater than 15° from the vertical) from entering the motor directly or through splashing or bouncing. The openings can also be specified with screens to prevent the entrance of foreign objects, and the motor can be specified either semiguarded or fully guarded.

Splashproof motors are similar to dripppoof except the ventilating openings prevent entrance of rain or objects falling at not greater than 160° from the vertical.

Externally-ventilated motors (force-ventilated) have cooling provided by a motor-driven blower mounted on the machine enclosure. Guarding also is available.

Pipe-ventilated motors have ventilating inlet openings arranged for attaching pipes or ducts that carry cooling air from some external source.

Weather-protected, Type I motors (WP-I) have ventilating openings constructed to minimize the entrance of air-borne water, snow,or particles.

Weather-protected, Type II motors (WP-II) have intake and discharge passages arranged so that airborne particles blown into the machine can be discharged without entering the internal passages leading to electrical parts. The path of the air within the motor includes three abrupt changes in direction, each being not less than 90°. Additionally, the intake air path includes an area of low air velocity not exceeding 600 fpm to minimize the possibility of moisture or dirt being carried into the electrical parts.

Encapsulated motors are not a separate type of open motor, but rather a method of winding treatment. Where an open motor is needed for a very high humidity application, the windings can be encapsulated usually in an epoxy-type resin. Usually, these motors will also have all external parts treated for protection against corrosion.

Totally-enclosed, nonventilated motors (TENV) have no openings for entrance or exit of cooling air and no provisions for external cooling.

Totally-enclosed, fan-cooled motors (TEFC) provide exterior cooling by an integral fan mounted usually within a shrouded enclosure on the end of the motor and driven by the motor shaft. The motor can be specified with a guarded fan enclosure that prevents passage of a rod or other probes.

Motor controllers can be specified grouped into motor control centers containing electromechanical starters, PLCs, and variable-speed drives.

- ❑ Type
  - ❑ Squirrel-cage
  - ❑ Wound rotor
  - ❑ Other
- ❑ Ratings
  - ❑ Horsepower
  - ❑ Voltage
  - ❑ Full-load amperes
  - ❑ Number of phases
  - ❑ Frequency
  - ❑ Speed
  - ❑ Service factor
  - ❑ NEMA design
  - ❑ Code letter
- ❑ Insulation class
- ❑ Ambient temperature
- ❑ Altitude
- ❑ Duty
  - ❑ Continuous
  - ❑ Intermittent
- ❑ Efficiency
- ❑ Enclosure type
- ❑ Frame size
- ❑ Construction
  - ❑ Horizontal
  - ❑ Vertical
  - ❑ Inline
  - ❑ Other
- ❑ Terminal box
  - ❑ Location
  - ❑ Size
- ❑ Shaft length
- ❑ Accessories
  - ❑ Space heaters
  - ❑ Temperature detectors
  - ❑ Other

Fig. 6.1. Check list for induction motors.

Totally-enclosed, pipe-ventilated motors (TEPV) have inlet and outlet openings arranged for the connection of ducts or pipes for conveying cooling air. Totally enclosed motors in which cooling is provided by a heat exchanger include water-to-air (TEWAC), and air-to-air cooled (TEAAC).

Explosionproof motors can withstand an internal explosion of hazardous gas or vapor without communicating hot gases, sparks, or other material to the outside hazardous atmosphere where they could touch off another explosion. These motors are marked to show the class, group, and temperature for which they are approved. The temperature must not exceed the ignition temperature of the specific gas or vapor that will be encountered.

MISCELLANEOUS FEATURES

The following specific construction features of a motor must be included in a specification.

- Horizontal or vertical mounting.
- Special shaft requirements.
- Motor junction-box location; standard is on the right when looking from the side opposite the shaft.
- Space heaters.
- Embedded temperature detectors in the windings.
- Frame size when needed to replace and existing motor.

Fig. 6.1 is a check list of items that must be included in specifications for induction motors. Similar ones can be developed for DC, synchronous, and other type motors.

MOTOR CONTROLLERS AND DRIVES

A motor controller consists of a power circuit, a control circuit, and motor protective devices. In electromechanical motor starters, these functions are performed by contactors, relays, thermal sensors, etc. In solid-state starters, they are performed by power semiconductors in the power circuit, while integrated circuits and microprocessors chips provide control and protective functions. Specifying starting equipment is dependent upon the available power supply, installed electrical system, and the load characteristics.

Manual starters. Fractional-hp motors generally require manual starters that provide an ON/OFF control as well as overload protection. When overload protection is provided integral to the motor, a hp-rated switch is suitable. Short-circuit protection in this case is provided by the branch-circuit protective device (fuse or circuit breaker) in a panelboard. When the fractional-hp motor must be started remotely, then a single-phase magnetic starter can be used.

Electromechanical starters. Across-the-line versus reduced-voltage

- ❑ **Short-circuit protection**
  - ❑ Available short-circuit current
  - ❑ Molded-case circuit breaker
    - ❑ Thermal-magnetic trips
    - ❑ Magnetic-only trip
    - ❑ Solid-state trips
    - ❑ CB and fuses
    - ❑ Fused disconnect
    - ❑ Relays
- ❑ **Starter**
  - ❑ Type
    - ❑ Manual
    - ❑ Across-the-line
    - ❑ Reduced voltage
    - ❑ Two-speed
    - ❑ Reversing
    - ❑ Solid-state
    - ❑ Other
  - ❑ Rating
    - ❑ NEMA
    - ❑ IEC
  - ❑ Size
  - ❑ Voltage
  - ❑ Hertz
  - ❑ Auxiliary contacts
  - ❑ Control transformer
    - ❑ Size
    - ❑ Protection
- ❑ **Overload protection**
  - ❑ Relay
    - ❑ Melting-alloy
    - ❑ Bimetallic
    - ❑ Solid-state
  - ❑ Embedded detectors
  - ❑ Electronic
- ❑ **Enclosure**
  - ❑ Individual
  - ❑ Grouped
  - ❑ NEMA type

Fig. 6.2. Check list for controllers of AC squirrel-cage induction motors.

starting is a major decision facing the specifier of most induction motors. The distribution system must be capable of supplying starting current without inducing excessive voltage drop on the line. Torque sufficient to accelerate the load is dependent upon applied voltage. Full-voltage (across-the-line) starting

For Outdoor Use

Provides a Degree of Protection Against the Following Environmental Conditions	Enclosure Type Number 3	3R	3S	4	4X	6	6P
Incidental contact with the enclosed equipment	X	X	X	X	X	X	X
Rain, snow and sleet	X	X	X	X	X	X	X
Sleet*	–	–	X	–	–	–	–
Windblown dust	X	–	X	X	X	X	X
Hosedown	–	–	–	X	X	X	X
Corrosive agents	–	–	–	–	X	–	X
Occasional temporary submersion	–	–	–	–	–	X	X
Occasional prolonged submersion	–	–	–	–	–	–	X

*Mechanism shall be operable when ice covered

For Indoor Use

Provides a Degree of Protection Against the Following Environmental Conditions	Enclosure Type Number 1	2	4	4X	5	6	6P	12	12K	13
Incidental contact with the enclosed equipment	X	X	X	X	X	X	X	X	X	X
Falling dirt	X	X	X	X	X	X	X	X	X	X
Falling liquids and light splashing	–	X	X	X	X	X	X	X	X	X
Circulating dust, lint, fibers and flyings	–	–	X	X	–	X	X	X	X	X
Settling airborne dust, lint, fibers and flyings	–	–	X	X	X	X	X	X	X	X
Hosedown and splashing water	–	–	X	X	–	X	X	–	–	–
Oil and coolant seepage	–	–	–	–	–	–	–	X	X	X
Oil or coolant spraying and splashing	–	–	–	–	–	–	–	–	–	X
Corrosive agents	–	–	–	X	–	–	X	–	–	–
Occasional temporary submersion	–	–	–	–	–	X	X	–	–	–
Occasional prolonged submersion	–	–	–	–	–	–	X	–	–	–

Table 6.2. Recommended applications of enclosures for motor controls (derived from NEC Table 430-91).

should be specified unless there are limitations on inrush currents or if the load is a high-breakaway type.

The decision between a mechanical and a solid-state starter becomes important when reduced-voltage starting is required. Reduced-voltage starting provides a cushioned start that lowers starting currents and also has the effect of stabilizing the line voltage. There are several electromechanical methods used to obtain reduced-voltage starting. They include: part-winding, reactor or resistor, autotransformer, and wye-delta.

Solid-state starters inherently assure that there is no sudden application of power to the motor or abrupt transition from reduced voltage to full voltage. Many have, as a low-cost option, a "power-factor corrector" that permits significant energy savings, particularly for motors operated at very low loading, up to about 50 to 69% of full load.

Specialty starters. There are additional ways of starting induction motors, including the use of reversing starters, 2-speed starters, and others. Well pumps and fire pumps require equipment packaged to meet their special requirements. Synchronous motors also require specialized starting equipment.

Drives. Controllers for DC motors and speed-regulated AC motors are called "drives." They are complex electronic/solid-state packages that must be carefully specified to meet application requirements.

OVERLOAD PROTECTION

The most common electromechanical motor overload protection is furnished by the melting-alloy or bimetallic thermal overload relay. There are three standard types of heaters available:

- Class 10 must trip within 10 sec at 600% motor full-load current;
- Class 20 must trip within 20 sec;
- Class 30 must trip within 30 sec at 600% motor FL current.

The proper one that should be specified is dependent upon protective-device coordination curves for the system. The specifications should require that pertinent time-current characteristic curves be submitted. If the location of the device is to be in an area with an elevated ambient temperature, ambient compensation should be specified to avoid nuisance tripping.

Electronic overload relays provide protection for the more critical motors in a facility. They offer a broad range of protection such as over/under voltage, single-phase detection, ground fault, reverse direction, and others.

Other NEC recognized means of providing overload protection include: embedded winding temperature switches, resistance temperature detectors (RTDs) with remote protective relays, and thermistors with relays mounted on the motor. Selection depends upon the hp and degree to which the motor is critical to the operation, and the nature of the overall protection scheme for the particular project.

Fig. 6.2 is a checklist of specification requirements for a motor controller used with an AC squirrel-cage induction motor.

ENCLOSURES

NEMA-standard enclosures are used for most motor controllers, sensors, and other types of electrical control equipment. The enclosure types are based on the location and atmosphere in which the units are to be applied. When specifying motor control equipment, a reference (by NEMA Type) must be made in the specification to the type of enclosure desired. The classifications that are available are listed below. IEC-standard motor control equipment and control devices have equivalent enclosures.

- Type 1 is general purpose.
- Type 2 is dripproof.
- Type 3 is dusttight, raintight, and sleet/ice-resistant.
- Type 3R is dust-tight, raintight, and sleet/iceproof.
- Type 4 is watertight and dusttight.
- Type 4X is watertight and corrosion-resistant.
- Type 6 is submersible.
- Type 7 is Class I hazardous, air break.
- Type 8 is Class I hazardous, oil immersed.
- Type 9 is Class II hazardous.
- Type 10 is Bureau of Mines approved.
- Type 12 is industrial dusttight and driptight.
- Type 13 is "oiltight," mainly applied to control devices.

Table 6.2 shows the NEC recommendations for application of enclosures in indoor and outdoor locations.

When motor control equipment is to be installed outdoors, or in an area that will experience temperatures below 32°F, space heaters or equivalent means can be specified to prevent condensation within the enclosure.

CONTROLS

When a motor is part of a complex automatic type of equipment, process, or production line, control systems external from the starter or drive provide the means for integrating the motor controller with the other elements of the system. The following are typical control items.

Sensors. Limit switches, pressure, temperature, flow, and other similar devices tell the logic (brains) of a control system what is happening within a system. The devices are specified to meet the action or parameters that are to be measured.

Limit switches usually indicate that a mechanical device has moved from or has reached a specific location. Various types of actuator heads are available (lever arm, pushtype, "wobble stick," etc.) to signal the event to the logic. Lever-arm types can be spring-return types, one-direction spring return, gravity return, or maintained. Specialized types are utilized for cord-pull emergency stops on conveyor belts. Enclosures can be NEMA 1, NEMA 4, NEMA 7-9), or NEMA 13. From one to four circuit contact blocks can be

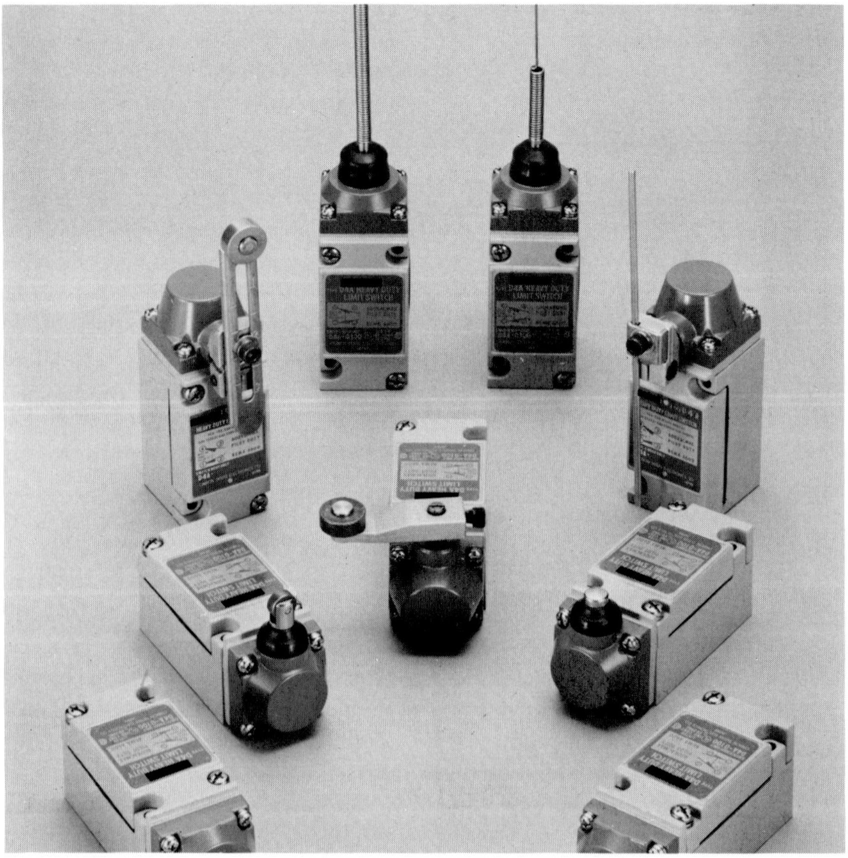

Sensors such as these limit switches come in a huge array of styles to meet the needs of the application. Specification of these items requires in-depth understanding of the control logic.

Pushbuttons, pilot lights and other pilot devices are essential compontents of control systems. They must be carefully specified, taking the system operator needs into account.

specified, with contact operation depending upon direction or other factors. The contacts can be in air, in a hermetically sealed tube, or solid state with direct compatibility with computer-type equipment.

Other types of sensors including photoelectric switches and proximity sensors have many variations that must be carefully analyzed and specified.

Pilot devices serve as interfaces between operating/maintenance personnel and the control system. The most common devices are pushbuttons, selector switches, and rotary cam switches. Pilot lights with nameplates marked START, STOP, RAISE, RUNNING, and similar markings of conditions that are desired or exist, assist in keeping the operator informed and in control of the production equipment.

As with limit switches, a seemingly simple element such as a pushbutton has many variations. The size, type, and color of the operator; number of contacts and their operating sequence; whether they are regular duty or oiltight; whether open or in an enclosure; nameplate size and inscription; and a myriad of other options, such as illuminated and push-to-test, are available. Other pilot devices have a similar diversity.

Logic devices receive the inputs provided by sensors and pilot devices and make decisions based upon the operating sequence dictated by the schedule for which it has been wired or has been programmed.

The most popular industrial/commercial logic device is the programmable logic controller (PLC). It is available in sizes ranging from the equivalent of a few relays, to systems accommodating thousands of inputs an outputs (I/Os). The PLC can be self-contained or its logic section can be in a centralized location, and the I/Os mounted in racks distributed at various locations near the equipment to be controlled. Outputs can be "powered" or "dry" and inputs can be specified to accept digital (ON/OFF), analog, or other signals, handle sophisticated mathematics, control PID loops, and many other variations. Voltage levels can range from 5VDC to 120VAC. Communication cards permit interchange of information between I/O racks and the logic section, as well as with other I/Os, supervisory computers, and other equipment.

Other electrical control equipment that perform a logic function include relays, timers, time-delays, drum programmers, computer-based distributed control systems, and instrumentation. Each has its full line of variations.

Because logic devices are so critical to the operation of a control scheme, the amount of details that need to be specified are greater than for others items. More often than not, it is selection of the logic system that dictates the choice of other control system components. A means of communications between the components (LAN, hard wire, coax, multiplexer, etc.) are often dependent upon the logic system chosen.

Output devices. Once the logic has made decisions based upon the inputs it has received, it energizes or deenergizes devices such as motor starters, solenoid valves, motor-operated valves, servo systems, or control circuits of other equipment. Voltage

Programmable logic controllers, relays, timers, and similar devices are the "brains" that accept input information from sensors, matches the data against its program, and determine whether motors that are part of complex systems should be started or stopped. Understanding the nature of the loads the motors are driving is needed in the specification process for these items.

levels, enclosures, and actuating motion must be considered and specified for these items.

Monitoring devices and alarms keep operating personnel aware of critical situations and equipment conditions.

Annunciators are often integrated into main control boards of centralized control rooms. These package units alarm conditions that are out of their set limits. Different sequences are available to distinguish the first event that caused a series of alarm conditions; to reset automatically after the alarm condition has returned to normal; for testing the lamps; and many more. Annunciators can be electromechanical or solid state and come in variations of alarm window configurations and legends. All must be specified as part of the overall control scheme.

Closely associated with annunciators are alarm horns, gongs, flashers, and similar devices. This same equipment is also used in independent control schemes to warn or alert personnel. Specification depends upon operating voltage, enclosure, sound level, etc.

Monitoring the process and allowing operators to control a process from a central location greatly increases efficiency of production.

Feedback devices such as speed switches are other items that must be specified to integrate into the control system.

An electrical installation sometimes requires specifying other than electrical equipment such as here where air is used to purge a starter enclosure.

INSTALLATION

7

ALTHOUGH the AC induction motor is the workhorse of the industry, installation procedures for all types of AC and DC motors, large and small, standard and special, low or high voltage are quite similar. There are also recommended practices that apply to the associated starting and control equipment of motors.

MOTORS

Receiving. When a motor reaches the jobsite, it should be thoroughly inspected for scratches, dents, or other signs of damage. This inspection should be done before the motor is moved from the shipper's truck or other vehicle. Check the motor nameplate for proper voltage, phase, frequency, horsepower, etc. Examine all literature provided with the motor and file with specifications and drawings for reference during installation and for guidance during startup and operation. If the motor is to be stored or installed in a damp location and built-in space heaters are not already supplied, installing a single-phase, low-voltage supply to the windings to combat moisture will allow the motor to reliably be put into operation when needed.

Installing motors and their driven equipment is one of the major parts of a large project. If the motor is delivered to the jobsite separately, normally it is the electrical contractor who will receive, check, and store it until it is ready to be mated to the driven equipment. Once the equipment is shimmed, leveled, and grouted in place, the electrical contractor assumes full responsibility for providing the power to the unit, terminating the cable, lubrication, and "bumping" the motor.

Handling. Transportation and rigging of large motors at the jobsite should be supervised by qualified personnel with experience in rigging and the use of heavy equipment.

Visual checks. Before a motor is transported from a jobsite storage area to the installation site, it is a good idea to make the following inspections.

- Motor enclosures to assure that the

enclosure matches the expected environment.

• Space availability at site for maintenance or replacement. Open motors having commutators or collector rings must be located or protected so that sparks cannot reach adjacent combustible material.

• Suitable guards where exposed current-carrying parts of motors will be subjected to dripping or spraying of oil, water, or other liquid, unless the motor enclosure is specifically for those conditions.

• Hazardous location motors are marked with an identification number that indicates the operating temperature range. The installer should check with the construction drawings and specs to make sure that this temperature will not be exceeded in the completed facility.

Lubrication. Before the motor is installed, small motors should have their shaft turned by hand to make sure it turns freely. If the motor is equipped with antifriction bearings, they will normally be prelubricated and ready for operation. However, large motors having sleeve bearings are usually shipped without lubricating oil in the bearings; often they are filled with an antirust fluid. If a drain plug is provided in the end bracket or bell, it should be removed prior to installation to drain any accumulated condensation. Bearings should be inspected through the sight-glass and bearing-drain openings for any accumulations of moisture and traces of oxidation, which should be removed. Then, fill the bearing reservoirs to normal level with the recommended high-grade industrial lubricating oil.

Mounting. The requirement for a level base prior to installation is critical. There are four points of mounting, one at each corner of the mounting base. All mounting points must be on the exact same plane or the equipment will not be level.

Before the concrete foundation is poured, locate foundation bolts by use of a template and provide secure anchorage (not rigid). A fabricated steel base between motor feet and foundation aids stable, permanent mounting. See certified drawings of motor, base, and driven unit for exact location of foundation bolts.

For smaller motors, sliding bases and adapters are available for use with T-frame motors when they are being installed to replace an old motor. Also, check whether other components or equipment such as gears, special couplings, and pumps are to be mounted on the motor. If so, be sure space is available.

After the motor base is in place and before it is fastened, shim to level. Use a spirit level (check two directions at 90°) to ensure that motor feet will be in one plane (base not warped) when base bolts are tightened. Set the motor on the base, install nuts, and tighten. Do not make a final tightening until after alignment.

Hazardous location motors have machined surfaces having very close tolerances. Extreme care must be taken during the installation process since any nicks or burrs may destroy the explosionproof or dust-ignitionproof features of the machine. If these mating surfaces are altered in any way, the machine will no longer be properly classified as a labeled motor.

Alignment. With few exceptions, a flexible coupling is typically used to connect the motor to the load. This type of coupling tolerates some misalignment; however, misalignment can cause vibration and/or stress on the motor bearings. Consequently, the shafts in all coupled applications should receive an installation lineup with a high degree of accuracy. A coupling should not be installed by hammering or pressing. Always heat the coupling to install it on the shaft.

There are several important steps to follow in attaining correct alignment of direct-connected drives.

• Position the motor on its foundation to obtain the correct spacing between the motor shaft and the driven shaft. This distance is specified by the coupling manufacturer.

• In the case of sleeve-bearing motors, positioning should limit the axial movement of the coupling to keep the motor bearing floating off the thrust shoulders. These bearings will not take continuous thrust. When positioning the shaft of motors with end play, the shaft should be placed at the midpoint of the end play. (Ignore the magnetic center indication.)

• Adjust the position of the motor by the use of jacksleeves, shimming, etc., until the angular and parallel misalignment between the two shafts is within the recommended limits as measured with a dial indicator with the motor bolted down. When adjusting the position of the motor, care should be taken to assure that each foot of the motor is shimmed before the motor is bolted down so that no more than a .002-in. feeder gauge can be inserted in the shim pack.

• It is recommended that "floating shaft couplings" or "spacer couplings" be installed on motors where the coupling alignment cannot be accurately checked and/or maintained. Misalignments of several thousandths of an inch will result when there are relatively small changes in the temperature differences between large motors and its driven equipment.

• After alignment with the load, bolt the motor in place with maximum-size bolts. It is advisable to provide for some variation in the location of the foundation bolts. This can be done by locating the bolts in steel pipe embedded in the foundation.

Electrical connections. When connecting the motor terminals to line leads, there are some specific recommendations that should be followed.

• Motor terminal box must be turned so that conduit or cable entrance puts the least strain on the raceway and cable terminations. For oversize conduit boxes such as those required for stress cones or surge-protection equipment, the mounting height of the motor and driven equipment may have to be increased for accessibility.

• Connector lugs must be sized to the conductor.

• Torque specifications must be adhered to in order to assure as high degree of reliability is possible. The terminations should be made up with a torque wrench.

Testing. After installation is completed, but before the motor is energized, make the following checks while the motor is disconnected from the driven equipment.

This DC motor actuator is being wired to its control contactor prior to its being mounted on a 24-ft high motor-operated valve of a pipeline.

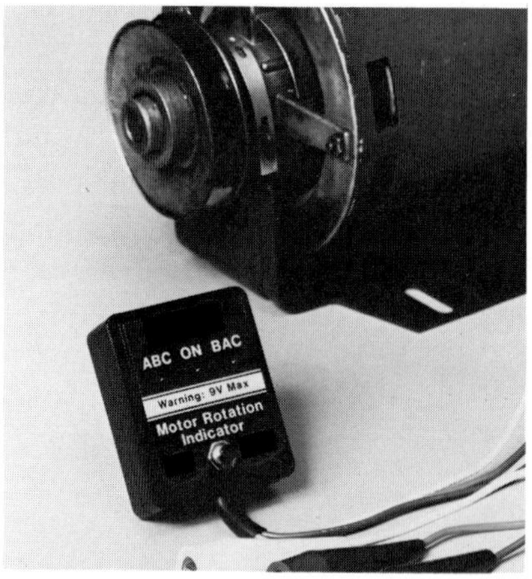

Phase rotation must be checked following the installation of an AC motor so that its load will be rotated in the correct direction when the motor is energized. Simple instruments are available to do the test. Interchanging any two of the motor leads of a 3-phase AC motor will reverse the direction of rotation.

• Connections of motor, starting, and control devices agree with wiring diagrams.

• Voltage, phase, and frequency of line agree with the motor nameplate.

• Service records and tags accompanying the motor. Be certain bearings have been properly lubricated and oil wells are filled.

• Perform insulation-resistance tests. For recommended minimum insulation resistance for the stator winding, see IEEE Publication No. 43, "Recommended Practice for Testing Insulation Resistance of Rotating Machines" for more complete information. If the insulation resistance is lower than this value, it is advisable to check for and eliminate moisture in accordance with accepted procedures.

• Rotate the shaft by hand. Be sure that the rotor turns freely.

• Start the motor at no load and let it run long enough to listen and feel for excessive noise or vibration

• Check direction of rotation. If incorrect, interchange any two line leads to reverse rotation on 3-phase motors.

• Record for future reference all pertinent data such as FL amps, volts, starting current, insulation resistance, noise level, and temperature.

• The equipment should be given a test run to verify that it gives satisfactory performance. Once performance has been verified, the machines should be doweled to their bedplates. Recommended doweling is two dowels per machine, one in each of the diagonally opposite feet, with the size of the dowels approximately one half the diameter of the hold-down bolts.

MOTOR CONTROL CENTERS

Proper installation of motor control centers (MCCs) speeds assembly, reduces labor costs, boosts reliability, and lengthens life of the equipment. Thus, MCC installation should be carefully planned well before the equipment is delivered. For example, the location in which it is to be installed should be well ventilated, free from excessive humidity, dust, and dirt. The temperature in the area should be between 0°C and 40°C (32°F to 104°F). Adequate access space (minimum 3 ft) in front of the MCC must be allowed for maintenance or changes. When two MCCs are installed back-to-back, the same required space must be provided on both front sides of the equipment. If the equipment is to be placed along a wall, a space should be allowed between the back of the MCC and the wall (6 in. for humid locations). MCCs should always be installed on a smooth, level foundation or pad.

Receiving. When delivered, inspect the MCC for any shipping damage immediately. Generally, any claims for shipping damage must be made against the carrier within a specified time limit. In most instances, the visual inspection will require removing plastic wrapping or crating. However, do not discard the material because it should be used to protect the units if the equipment will be in storage for some time before it is installed. The MCC should be stored in a clean, dry, ventilated location free from temperature extremes and specifically, not outdoors.

Handling. Use extreme care when moving and handling MCCs because a single shipping block of a typical unit (three to four vertical sections) will often weigh over 2000 lbs. An MCC subjected to rough handling should be inspected by the manufacturer for possible damage.

Some manufacturers ship their equipment on its side, while others ship the sections standing upright. Either technique is acceptable if properly braced and protected. The best way to handle an MCC is with a forklift or an overhead crane.

Motor control centers tend to be top heavy, and care must be taken to avoid

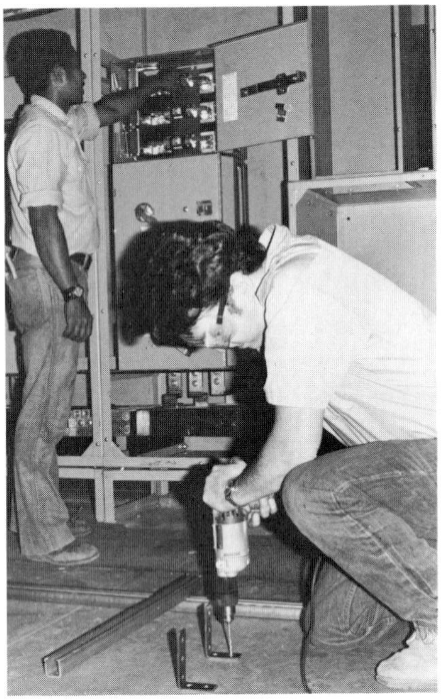

When installing an MCC, it is important that the unit be set on a level floor or pad, and that the MCC is firmly anchored to the floor.

Connecting the motor and control leads to the appropriate starter in a MCC requires attention to detail. Good connections are essential to the correct operation of the systems.

tipping over the equipment because this usually means a return trip to the factory.

Installation. The installation should always be done only by trained, experienced electricians who are familiar with electrical equipment. Observe all safety regulations. Never attempt to install an MCC section on an energized system. Establish a sequence in which work is to be done.

• Install any conduit that is to enter the bottom of the MCC, checking the manufacturer's drawings for proper conduit location. In some instances, a plate or template is provided for guidance.

• Pour the pad, if required, and imbed mounting channels if supplied.

• Adequately sized anchors must be embedded before the MCC is set into place.

An alternate is to use cinch anchors that are inserted after the MCC is in position. Usually, templates or drawings are supplied for proper positioning of the anchor bolts.

• Bolt the shipping blocks (section groups) together with the hardware supplied, and secure the sections to the mounting surface or floor.

• Splice the power busbars between shipping blocks. It is important that this be done after the sections are bolted together and secured to the floor; otherwise, the main bus insulators could be damaged.

• Make the ground bus splice connection.

• Install conduit. The top and bottom plates provided, normally can be removed to simplify the cutting of conduit entry holes.

• Power and control cables can then be pulled, routed to the proper vertical wireway, and secured to their terminations following the manufacturer's torque requirements. Be sure that all power conductors are adequately supported to withstand the available fault current of the system.

• Control wiring required between units in the MCC can be run in the vertical and horizontal wire troughs provided. If a Class II MCC is specified, much of this interwiring will have already been installed by the manufacturer.

Operational check. Once the installation is complete, an operational check should be made on the equipment. The following checklist will help to carry out an inspection after assembly or service.

• Remove any tools or debris left from installation.

• Double-check for any shipping damage and be sure that all equipment is clean and in good condition.

• Perform insulation-resistance tests on power wiring and buswork.

• If the MCC main protective device or any motor starters are furnished with ground-fault protection, it should be properly adjusted before energizing.

• Remove all blocks or other temporary holding means.

• Current transformers should have the secondary shunt bar removed. Do not operate a current transformer with its secondary open-circuited.

• Manually operate all switches, circuit breakers, and other mechanisms to make sure they are properly aligned and operate freely.

• Electrically exercise all switches, circuit breakers, and other mechanisms (but not under load) to determine that the devices operate properly. An auxil-

A visual inspection is needed before the MCC is energized. Check that wiring to terminal strips matches that on drawing, wires are securely fastened, and wires are identified.

Matching the enclosure to the environment is essential when applying individual combination motor starters within operating areas subject to moisture, dirt, or corrosive fumes or liquids.

iary source of control power may be required.

• Timers should be checked for the proper interval and contact operation.

• Check overload heater selection tables against motor full-load current to ensure that proper heaters or thermal overload units have been installed.

• Make a visual check that all load and remote-control wiring connections have been made to the proper terminal block and that they agree with wiring diagrams provided.

• Locate and check that all ground connections have been made.

• Inspect and check trip ratings of all circuit breakers, relays, and fuses.

• Install covers, close doors, and make certain they are properly tightened.

Energizing. Only qualified personnel should energize the equipment. If faults caused by damage or poor installation practices have not been detected during the checkout procedure, serious damage and/or personal injury can result when the power is applied. To minimize risk, there should be no load on the MCC when it is energized. Turn off all downstream loads, including distribution equipment and other devices that are remote from the MCC.

• Equipment should be energized in sequence. Start at the source end of the system.

• Energize the feeder to the MCC.

• Be sure all compartment doors are securely closed, and then close the MCC main circuit breaker (if one is provided).

• Close motor starter and branch circuit disconnect devices. During this procedure, be sure all barriers (if applicable) are in place and unit doors are closed and latched.

• Loads such as lighting circuits, starters, contactors, heaters, and motors may be energized.

INDIVIDUAL CONTROLLERS

Installation procedures for individual electromechanical or solid-state starters are similar to procedures for MCCs. Most starters of either type come in enclosures for wall or rack mounting. Racks can be fabricated by the installer from steel angles by welding, or slotted steel struts can be assembled into rigid supporting racks.

Enclosures. Make sure that the enclosure selected is suitable for the field conditions encountered. Also, check the location and environment. Excessive heat is the biggest enemy of a solid-state starter; dirt and humidity are enemies of electromagnetic starters. Always leave ample working space around the equipment.

Location. Be sure not to indiscriminately relocate the starter for convenience without checking that the motor disconnecting means is "in sight from" the motor location. This applies not only to individual motor disconnects but also to combination starters that incorporate the motor disconnect in the same enclosure with the motor starter. Note that "in sight from" means that one piece of equipment is visible from the other and is not more than 50 ft away.

CONTROL DEVICES

A large amount of the electrical equipment that is installed by the electrical contractor or facilities maintenance personnel involves control elements such as programmable controllers, pushbuttons, pilot lights, limit switches, photoelectric sensors, solenoid valves, and the like. The techniques employed for physically mounting these items are reasonably similar to those for other electrical equipment. However, more precision and attention to detail is required.

More than any other electrical equipment, control devices that are spread out over a wide area need to be integrated into a single system. For the installer, it is analogous to buying busbar, overcurrent devices, indicating lights, enclosures, etc., and having to assemble them to create a piece of switchgear. A great deal of thought and skill is required to accomplish the task in a workmanlike manner.

Location and mounting. Programmable logic controllers (PLCs) and most sensors and pilot devices are intended for use in harsh environments and are available in enclosures that meet any requirements. However, the installer should check that these items are suited for the environment in which they are to be located.

Most sensors should be mounted on a bracket that is sturdy enough to maintain the item in a fixed position, yet have slots or other provisions for making fine adjustments of location when required. This will greatly simplify the checkout and fine-tuning of the finished control system.

Wiring. Single-conductors or multiconductor control cable (usually No. 14 or No. 16 AWG) is extensively used for wiring between 120V control components. Techniques in installation and termination of these conductors are similar to that for power cable.

Increasingly, however, coaxial and twisted-pair cable are being used as more of the control devices being employed include microcomputers and solid-state components. More attention must be given to the quality of the connections that are made on these conductors. Use of proper tools and techniques is required.

Fiberoptic cable that will be used increasingly in the future will demand even more attention to splicing and terminations.

Identification. Because of the large number of conductors that interconnect the various components of a control system, it is necessary that the individual wires be identified at all locations in which they terminate. Usually, a wire-numbering scheme is established on construction drawings. Standard practice requires that self-adhesive or clip-on numbering devices conforming to the scheme be attached to each of the conductors. This will greatly simplify checkout of the system and allow troubleshooting and modification at a later date.

Checkout. Since a typical control system is made up of many separate field-installed-and-wired control components, a much more thorough checkout of the finished system is required than is the case with the power system. A "ringout" of the deenergized system is usually carried out by the installer to ensure that continuity exists and that items are connected to the proper terminals according to the drawings.

A "simulated" test of the energized system is usually done in conjunction with the designers of the system. This is intended to catch miswiring that was not detected in the ringout, errors in the design, and changes that must be made to accommodate equipment added during the construction phase of the project. Such a check will greatly reduce the time required for the actual startup of the facility.

Sensors such as proximity switches, operator interface devices such as pushbuttons, logic elements such as relays, monitoring devices such as annunciators, and output devices such as solenoid valves make up a control system. Installation electricians require special skills and training.

MAINTENANCE

8

In modern facilities, unscheduled outages of the electrical system or long repair shutdowns are intolerable because of the resultant high cost of downtime that eats deeply into profits. The item that is most critical to keeping the system reliable is maintenance. Motors and their associated starting equipment and controls are among the most important of these items that must operate correctly. Knowing how to properly service this equipment aids greatly in extending its operating life.

PREVENTIVE MAINTENANCE

A well-planned preventive maintenance (PM) program is the key to dependable, long-life operation of electrical equipment. The purpose of scheduled PM is to find and replace any components that might be subject to failure and prevent good components from failing prematurely. A good maintenance program fulfills both of these qualifications and will greatly reduce the risk of unplanned downtime. This will help the facility run at maximum efficiency with minimum downtime.

Management sometimes resists the investment in the proper tools, instru-

Preventive maintenance is a technique that can be employed to limit unscheduled outages. In an MCC, it is often possible to deenergize a motor, disconnect the wiring, remove the entire pan and replace with a spare, reconnect the wiring, and put the motor back into operation in a very short period of time. The contactor, short-circuit protective device, and overload relay in the pan can then be cleaned, checked out, and refurbished if necessary, and then be used to replace another that is being taken out for servicing. This scheduled maintenance greatly reduces the possibility of a breakdown that interrupts production.

Lubrication is one of the essential maintenance tasks that must be performed on a motor and its load. As simple an item as loose belts can cause loss of efficiency and poor performance.

ments, practices, or technical assistance that would guarantee effective motor performance. Therefore, it is very important to be able to explain how a properly planned motor preventive maintenance (PM) program can be justified.

• List the advantages gained through the implementation of a motor maintenance program. Do this by collecting case histories of motor breakdowns and the cost of resultant lost production. Show how budgeted PM costs are significantly less than the cost of lost production.

• Organize and set up the budget for a motor PM program. It must be effective and, at the same time, its cost must be kept to a practical minimum. Don't underestimate; a PM program won't work if you don't have the proper test equipment and tools, and trained persons to properly apply them.

• Determine which procedures are essential and whether they should be performed by facility electricians or by a service organization geared to do the job.

• Select the best motor-maintenance techniques and determine to what extent they should be applied. For example, should you check for possible bearing trouble on a motor simply by feeling components for overtemperature and listening for unusual sounds, or should you install temperature monitoring devices and make inspections using a stethoscope or an infrared scanner?

BASIC MOTOR MAINTENANCE GUIDELINES

Manufacturers' literature usually contains detailed instructions for the maintenance of their motors. Unless a good filing system is kept, very often these guidelines are misplaced or discarded. However, there are certain generalized principles that meet the requirements of maintenance of most types of motors.

• Lubricate regularly according to manufacturer's instructions. On sleeve-bearing and other oil-lubricated machines, check oil reservoirs on a regular basis. In poor environments, change oil at least once a month. Never over-lubricate; excess grease or oil gets into windings and deteriorates insulation. Be sure to use only the lubricant specified for the machine in question. However, check into the use of modern lubricants that have excellent life and lubricating qualities.

• On essential motors, or those that are heavily used or frequently duty cycled, check bearings daily using a stethoscope or infrared scanner. Check bearing surface temperature with a thermometer, electronic temperature sensing devices, or stick-on temperature-indicating labels. Compare the temperature of hot bearings with the temperatures of normally operating bearings. Check oil rings and watch for excessive end play.

• Check air gap between the rotor and stator with feeler gages at least annually. Measurements should be made at the top, bottom, and on both sides of the stator. Differences in readings obtained from year to year indicate bearing wear.

• Check belt tension. Belts should have about 1 in. of play. Sheaves should be seated firmly with little or no play. Couplings should be tight and operate without excessive noise. An alignment check should be made on all motor-generator sets and on motor-load couplings when trouble is suspected. Methods of checking alignment with a dial indicator and a straight edge are shown in **Fig. 8.1**.

• Inspect brushes and commutators of DC motors for excessive wear. Check brushes for proper type, hardness, and conductivity as well as how they fit in brush holders. Check spring pressure of brushholders with a small scale. In most instances, pressure should be 2 lbs, or slightly more, per sq in. of brush cross-sectional area.

• Check mounting of motors regu-

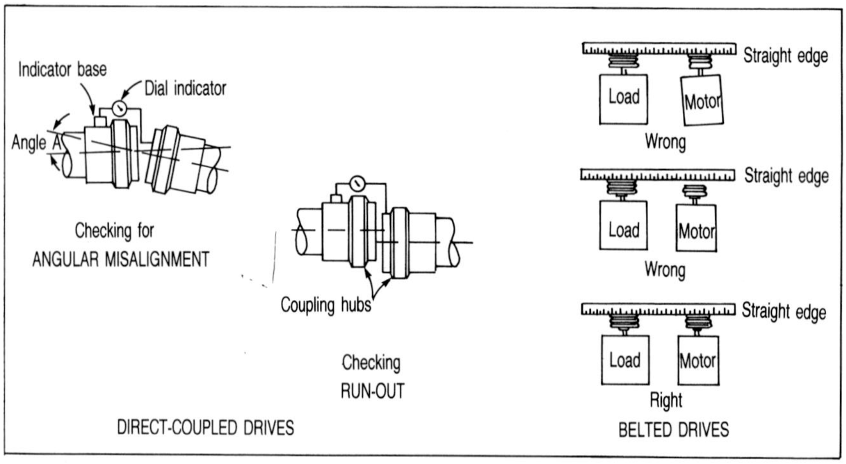

Fig. 8.1. Proper alignment between motor and load extends bearing life. On the left, dial indicator reveals angular misalignment of shafts. Proper alignment is indicated when angle A is zero. Next, run-out is checked with a dial indicator. This check reveals any small offset at coupling. A method for checking belted drives with a straight edge is shown at the right.

Cleaning the inside of an enclosure containing a motor starter and auxiliary control devices can ensure continued reliability.

same on all poles. Watch for frayed flexible leads.

• On essential controls, perform contact-resistance tests with a low-resistance ohmmeter on a regular basis. Proper contact resistance should be about 50 micro-ohms (50 millionths of 1 ohm). Record readings for future comparison. This will indicate trends in the condition of contacts.

• Overload relays should receive a thorough inspection and cleaning. Check for proper setting and recheck that the rating or trip setting takes into account ambient temperature as well as the higher inrush currents of modern, energy-efficient motors. To insure reliable operation, relays should be tested and calibrated every one to three years. Special equipment, such as an overload relay tester can be used to perform these tests.

larly. Check mounting bolts for tightness, steel base plates for possible warpage, and concrete base for cracking or spalling.

• Annually, perform vibration-analysis tests. Excessive vibration may be hard to detect, but detecting it is important because it could shorten motor life significantly.

• Keep motor clean and cool. In poor environments, blow out dirt with dry compressed air (no more than 50 lbs) as often as needed. In high-temperature locations, consider the use of energy-efficient motors that operate cooler than standard motors. Excessive ambient temperatures will shorten motor life.

• Pull and disassemble important motors during summer shutdowns. Thoroughly inspect, test, clean, and check bearings and couplings. Complete reconditioning of the motor may be required.

• Keep accurate records. Perform annual insulation-resistance and other appropriate tests. Important motors should also receive a thorough visual inspection, as well as voltage and current checks. All values should be recorded and compared each year. The trend of the readings will indicate the condition of the motor and offer a guide to its reliability.

MOTOR CONTROLLER MAINTENANCE

Like maintenance procedures for motors, those for motor control equipment are usually provided by the equipment manufacturer. Because the equipment in question is more varied and contains many more mechanical and/or electrical components, it requires more attention and expertise. Some general principles of maintenance, however, do apply to all such equipment.

• Keep control equipment clean. In poor environments, blow out dirt weekly, otherwise a quarterly or semi-annual cleaning should be adequate. Be sure dust or contaminants are kept off equipment. This is important because dust may contain conducting materials, which could form unwanted circuit paths resulting in current leakage or possible grounds or short circuits.

• Moving parts should operate easily without excessive friction. Check operation of contactors and relays by hand, feeling for any binding or sticking. Look for loose pins, bolts, or bearings.

• Check contacts for pitting and signs of overheating, such as discoloration of metal, charred insulation, or odor. Be sure contact pressure is adequate (see manufacturer's specs) and is the

MOTOR CONTROL CENTER MAINTENANCE

Any electromechanical device, including MCCs, begins to wear out as soon as it is put into service. If any equipment or part is going to fail, there is a good chance it will fail when needed most. This is not due to Murphy's Law; instead, equipment simply works hardest when it is needed the most. The harder the equipment is working, the greater the effects of heat, friction, and mechanical shock. To minimize unscheduled downtime, it is recommended that an electrical PM program be instituted.

To properly maintain an MCC, it should be deenergized. This means the production line must be shut down. There is often a reluctance to do so. However, downtime can be scheduled even in high-production facilities.

The general recommendation is that the MCC should be serviced once a year. However, unusual service conditions may necessitate performing maintenance more often. The amount of maintenance performed can also be varied to suit the particular application. However, the more extensive the maintenance program, the lower will be the risk of unscheduled downtime and lost production. A typical electrical preventive maintenance program is as follows.

• Deenergize all equipment prior to performing maintenance. Turn off power supplying the MCC and deenergize any control circuits that may be providing control power from other sources. Be sure any capacitors are totally discharged. All circuits should be locked out or tagged to be sure they are not inadvertently reenergized.

• Check the structure for corrosion. Touch up the paint where necessary.

• Inspect all busbars and incoming line compartments. Tighten, where required, the incoming cable lugs, bolted bus connections, and connectors on the main power bus. Replace any busbars or connectors that show damage from mechanical stress, heating, or arcing. Never try to repair a bus component by filing or sanding. Most bus components are plated. Any removal of this material may result in premature component failure.

• Inspect all splice connections and all connections between the vertical bus and the main power bus. All lugs and bolts on busbars should be torqued to the manufacturer's recommended torque values.

• Inspect all insulators, braces, and barriers. Clean off any accumulation of dust or debris. Do not use compressed air. Use clean rags or a vacuum cleaner. Replace any components that show signs of arcing, damage, tracking, or excessive heat.

• Inspect the individual controller units and the control devices they contain. Check each of these unit assembly connections to the main buses for signs of overheating. Check stabs for damage. If damaged they should be replaced.

• Check for proper ON and OFF operation of the operating mechanism of the circuit breaker or fused switch, as well as proper interlocking with the door.

• Manually trip and reset the circuit breaker and check for proper trip indication on the operating mechanism. A circuit breaker test set should also be used periodically to assure that the breaker is calibrated and operational. Check fuses manually for looseness to be sure that the fuse clips have not lost their tension due to overheating.

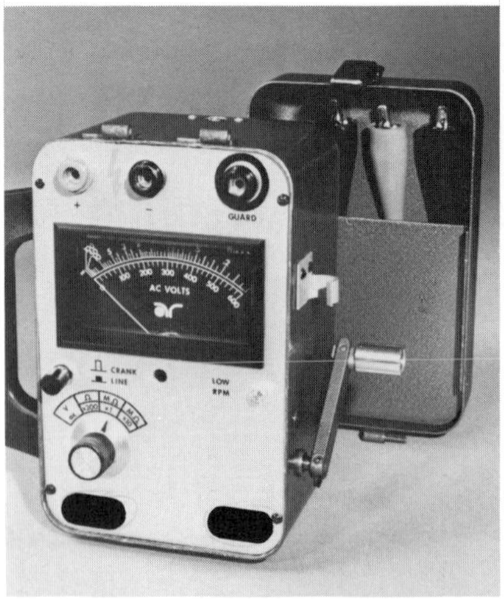

Insulation resistance check on the internal and external wiring on a periodic basis will pick up any deterioration in the insulation systems.

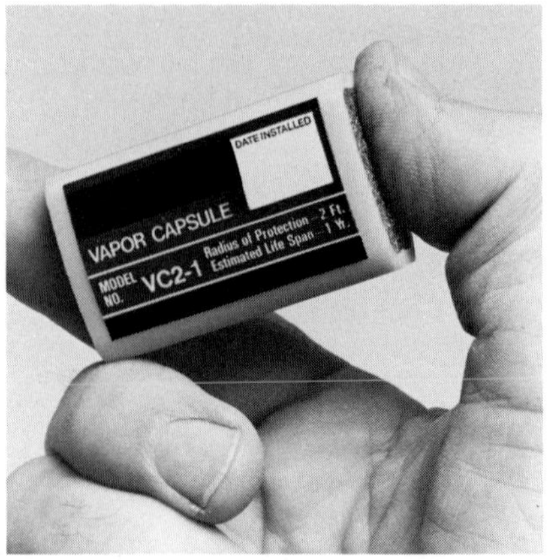

Inhibiting corrosion within a starter enclosure is important when it is mounted outdoors, in a wet location, or where subjected to corrosive atmospheres. Various types of solid or spray-on inhibitors can be effective.

• Inspect the contacts of each motor starter. Starter contact surfaces will normally wear away over a period of time and need to be replaced. Starter contacts will normally look somewhat rough and pitted. Do not dress the contacts with a file or sandpaper, as this only wastes contact material.

• Manually trip the overload relay to ensure proper operation. Check that the thermal unit or heater size matches the manufacturer's recommendations based on actual motor full-load current.

• Check for proper operation of contactors, relays, timers, and other control devices either by actual operation using control voltage, but with power components made inoperative, or by checking each with an appropriate test instrument.

• Check for the proper operation of interlocks.

• Inspect all power and control wires of the individual compartment pans, replacing any wire that has worn insulation or shows signs of deterioration.

• Before the equipment is returned to service, perform an insulation-resistance test to detect any short circuits in the system, as well as provide an indication of the condition of the bus insulation system and cable insulation. Record the readings from the tests for future comparisons.

PROTECTING AGAINST CORROSION

Electrical equipment in starters is often attacked by atmospheres containing salt or other chemicals that when combined with moisture result in corrosion. Shortened life and increased contact resistance result. Space heaters are often installed in enclosures and panels in an attempt to prevent a buildup of moisture. There is another approach that will help reduce the corrosion problem.

Corrosion-inhibiting compounds are increasingly being adapted to this purpose. One popular form is a capsule that contains a vapor-phase inhibitor. When the capsule is removed from its sealed package, it begins to emit an invisible, odorless, nontoxic vapor that is diffused throughout the surrounding atmosphere until the air is saturated. The vapor passivates the metal surfaces against atmospheric corrosion by reducing the electrochemical activity of the metal surfaces.

Capsules are particularly suitable for corrosion protection of enclosures holding electrical gear such as relays, contactors, switches, connectors, and electronic assemblies. The released vapor does not increase contact resistance, nor does it affect the characteristics of insulation that is part of the equipment or wiring.

Generally, corrosion inhibitor cap-

Modern rewinding techniques including vacuum pressure impregnation of windings allows upgrading of a motor, increasing its efficiency, and extending its useful life.

sules are marked with a radius of protection. The metal surfaces to be protected must be within the radius specified. Also, the location of the capsule plays a role in proper protection. Therefore, the size and shape of the enclosure must be considered. In some cases it may be advantageous to install more than one capsule within the enclosure. Overprotection by installing more capsules into an enclosure than are absolutely essential is not possible because the amount of protective vapor that the air can contain is limited by saturation, which is the same regardless of the number of capsules within the enclosure.

The life expectancy of a capsule is affected by the rate of evaporation of the chemical and by the rate at which the chemical decomposes. Life span is most often determined by conditions that accelerate the decomposition of the capsule, such as high temperatures (above 150°F) and aggressive atmospheres (salt, SO, etc.).

Vapor capsules work best within enclosures where air exchange with the outside atmosphere is limited. However, the enclosure does not have to be sealed or airtight. A rule-of-thumb is that up to two changes of air per day within an enclosure will not significantly reduce the effectiveness of the capsules. Another matter that must be considered when applying this type of corrosion protection is the need for proper sealing of the enclosure against the entrance of water. It is sometimes assumed that since the capsules are effective in preventing corrosion due to water vapor in the air, they will also protect against the effects of standing water. This assumption is incorrect. All enclosures should be sealed against water entrance or have adequate means of permitting the water to drain out.

Equipment in larger enclosures or open equipment can be similarly protected by spraying on corrosion-inhibiting chemical solutions.

REWINDING MOTORS FOR EFFICIENCY

Operating efficiency of rewound motors is important because energy costs have soared. To keep costs down, it is essential to know as much as possible about the factors affecting motor efficiency.

There are six components of losses that affect the efficiency of an AC induction motor: I^2R loss of the stator winding; I^2R loss of the rotor winding; eddy-current loss and hysteresis loss in the iron core; stray-load losses; windage and friction losses; and auxiliary items such as the nature of the load.

Any one or all of these losses could be increased during rewinding or rebuilding of a motor. Thus, it is essential that modern motor rewind methods be used to assure the retention or improvement of motor characteristics.

The worst rewind practice is burning out the motor at a temperature above 650°F to strip out the old winding, or permitting the old winding to ignite and rise above the oven setpoint. This can destroy the interlaminar insulation in the stator core and increase eddy-current losses.

It is also important to use the proper winding conductor sizes. If the cross-sectional area of the conductor is reduced, the resistance will go up and the I^2R losses will go up, with a corresponding decrease of efficiency and increase of operating costs. The use of a smaller diameter wire occurs when a motor rewinder attempts to shorten the time required for rewinding to reduce costs. If a smaller diameter wire is used, the wires can be inserted in the slots much more easily and quickly; but this is a detrimental practice.

Another bad practice is to drop the number of turns. This makes winding easier and faster; and although it is true that it reduces winding resistance, it increases the starting current, starting torque, and full-load torque and will increase stator-core loss due to increased flux densities.

Motor efficiency can be maintained or, in some cases, improved during rewind. One effective way is to increase the size of the stator conductors that are replaced during rewinding. The number of turns must not be changed, but sometimes enough extra space is available in the slots to permit use of a larger size wire. Some older motors have very thick, paper-like insulation in the slots, which can be replaced with a thinner insulation that has a higher dielectric and mechanical strength than the original to provide more slot space for copper.

Another way to reduce stator losses without changing turns is to wind with much shorter end turns. Also, any motor that was originally wound with aluminum wire can be greatly improved

Locating the source of a motor failure, such as a broken rotor bar, is difficult and normally requires special techniques and test equipment.

by a change to copper wire.

Hand winding by skilled people enables the slots to be packed much tighter and the coil length to be shortened so efficiency can be improved. This technique of rewinding takes more time for the engineering and the actual labor required, but the result is well worth the effort because of the real value gained in longer life, more dependability, increased efficiency, and lower operating costs.

Another recommended procedure is to have all motor stators screened by a core-loss test prior to stripping. This will identify motors with heavily damaged stator iron that will most likely require replacement. If it is a critical motor that is not immediately replaceable, it should have its iron repaired and reinsulated.

DETECTING DEFECTIVE ROTOR BARS

One of the toughest faults to find in a squirrel-cage induction motor is a defective rotor bar. Usually, a rotor-bar failure is found where the rotor bars are joined to the end-ring assembly. Typical failures found include: one or more bars open (broken loose from the end ring); partially open or cracked bars at the end ring; high-resistance joints (probably caused by a fine hairline crack); and voids in the end ring at the bar attachment.

Finding a totally open rotor bar is usually easy, but locating a partially open bar is tougher. In a typical case, a 350-hp medium-voltage motor ran for about two hrs at full load, then motor current would increase from its 44A full-load rating to a steady 47A, and the motor would start vibrating. Because this is a prime indication of a rotor-bar problem, the motor was sent to a motor repair shop.

The first step taken at the repair facility to locate the problem was to have the rotor sandblasted for a visual inspection. Nothing was spotted, so a basic growler test was conducted; this revealed no open bars. Next, a test was conducted passing 1000A at 15VAC through the shaft to excite the squirrel-cage winding. This again indicated that there were no open rotor bars. An infrared scan of the end ring was carried out but did not reveal any hot spots. An ultrasound test was attempted, but it was not definitive since the end ring had integral cast fans that created a highly irregular surface. The last test was with an oscilloscope and a special coil wound to fit around the rotor. A large growler was used to induce current into the squirrel cage. Still, there was not a clue as to where the problem might be, or how it could be revealed.

Thinking that perhaps some parts of the squirrel cage might move while running causing some components to open, the rotor was driven at 3000 rpm with no results. Next, to account for the effect of high temperatures, the rotor was heated to 300°F in an oven, and spun again at 3000 rpm; no faults were detected. In desperation, one end ring was sawed off and the problem was finally found. Large voids existed in the end-ring casting, probably as a result of improper manufacture. The rotor was rebuilt with aluminum bars and end rings and the problem was solved.

There is no question about the need of the industry for a test procedure that will positively and effectively reveal a defective (but not totally open) end ring. Tests that are made on rotors with casting voids and cracks before and after assembly will not normally show that the motor is defective.

There are a number of tests that sometimes detect the partially open rotor bar or end ring. Research is being done in an effort to solve this problem. In the meantime, vibration of motors can be checked at full load, and the motors also can be checked for excessive full-load current. At the moment, these tests remain the best indicators of rotor-bar problems.

STARTING HIGH-EFFICIENCY MOTORS

Combination starters are selected and coordinated to give safe and proper operation of contactor, overload relay, heater, and short-circuit protection. In particular, the motor circuit protector (MCP) was developed to give superior motor protection by providing faster tripping against low-level faults before they reach more destructive levels. Because it is a "magnetic-trip-only" device, the NEC classifies it in a category that limits its maxiumum trip setting to 1300% of motor full-load amperes. This restriction is significant when applying high-inrush motors, particularly high-efficiency motors with a lower-than-normal full-load ampere rating.

High-efficiency motors are by no means the only motors to experience this nuisance tripping problem, but are more likely to. This is more the case with the higher ratings, say above 75 hp, and with 2-pole designs in particular. The 4- and 8-pole designs are less critical.

Guidelines that are considered to be generally applicable have been developed when nuisance tripping becomes a problem.

- Review application circumstances (if possible) and starting current (Code letter) of all motors rated 100 hp and above.
- All high-efficiency motors as well as motors with Code H or higher locked-rotor values should be protected with thermal-magnetic (inverse-time) circuit breakers.
- The remaining motors should be examined in detail using the manufacturer's published data to determine if the design has a locked-rotor current greater than 600% of the full-load current. This will require some simple calculations. If the ratio exceeds 600%, a thermal-magnetic circuit breaker will probably be required.
- Other motors may be expected to operate satisfactorily with MCPs.

ISOLATORS FOR ROTATING EQUIPMENT

Shock and vibration, if not correctly compensated for, may shorten the life of rotating equipment. Motors experience hazardous vibrational forces caused by irregular operation, unbalanced dynamic loads, and worn or out-of-balance rotating parts. The source of damaging shock and vibration may come from external sources or from the equipment itself. The effects of both shock and vibration can be reduced or minimized with the proper use of isolators.

Isolation systems are used as an interface to protect sensitive equipment from the dynamic forces and motion of vibration and shock environments. They function to attenuate the impressed forces by virtue of their controlled elasticity and damping characteristics. The two most often specified isolators for rotating equipment applications are: elastomeric isolators, and helical cable isolators.

Elastomeric isolators are molded of natural or synthetic rubbers. Operation of this type of isolator primarily involves stressing the elastomer in either shear or compression. Elastomers have limitations in corrosive environments, temperature, and mechanical stress they can be subjected to. The approximate stable temperature extremes for most elastomer isolators is in the range of -50°F to +165°F. A wider temperature range may result in surface cracks under stress as well as changes in characteristics.

Under continuous stress reversals involving a high deflection magnitude, elastomers perform well initially but tend to drift in stress/strain characteristics over a period of time. To expect reliable operation, it is advisable to limit strain to approximately 20% in compression and to under 50% in shear.

Helical isolators are assemblies of stranded steel wire rope formed between metal retainers. Their large dynamic displacement attenuates heavy shocks, while their inherent damping capabilities enable them to absorb and dissipate large amounts of low- and/or high-frequency vibration. Cable is wound in a helical fashion between two metal bars to assure reliable shock and vibration control, regardless of the direction of applied force. As a base mount, even in cantilevered mounting, the helical isolator provides effective protection in compression, tension, shear, or roll. They are made of stainless steel or a combination of corrosion-resistant aluminum alloy and stainless wire rope. They function within large temperature extremes and resist ozone, oil, grease, sand, salt spray, and organic solvents.

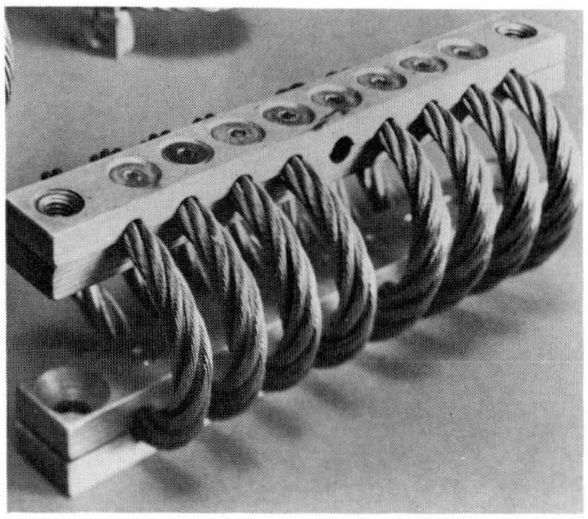

Vibration is a major enemy of electrical equipment, including motors. Large motor driven loads often can benefit by the installation of vibration dampers.

Following are items of information needed in recommending or designing an isolation system.

- Make and weight of the machine. This information is usually stamped on the nameplate or available in purchasing documentation.
- Forcing frequency at normal operating speeds. Forcing frequency is the major vibrational frequency of a machine expressed in Hz.
- Extent or range of speed variations, and how often they occur.
- Machine geometry, including its center of gravity, its "footprint" (its usual mounting points), and its space envelope (sway space).
- Machine's critical frequencies.
- Direction of oscillation.
- Presence of sources of external vibration, such as other machines resting on the same foundation.
- Applicable environmental conditions (ambient temperature, humidity range, corrosive conditions, etc.).

Installation of an isolation system is a straight forward job. Once installed, anticipated static deflection of the isolators should be checked and verified. Load should be equal on all mounts. Finally, after the rotating equipment has been operational for six months or so, any significant changes in shape, height, etc., should be investigated.

SERVICING WYE-DELTA STARTERS

Wye-delta starters are reduced-voltage units used with six-lead wye-delta induction motors. The size of the motor can be as small as 10 hp or as large as 1000 hp. Typically, they are used where the mechanical load is not driven while starting the motor. These starters are easy to maintain but often are neglected. A few minutes of periodic inspection can save a lot of headaches, especially when the starter operates a critical piece of equipment.

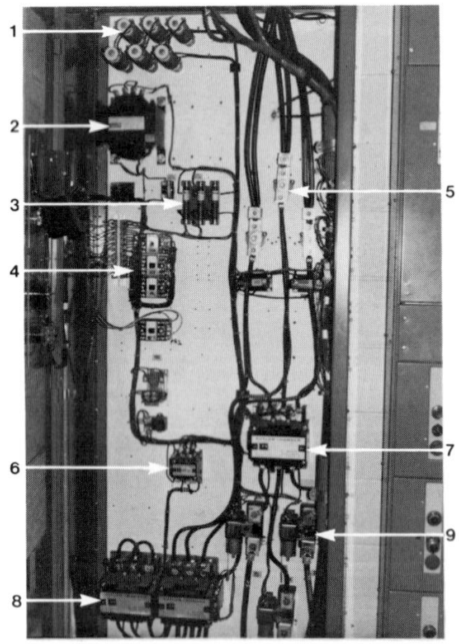

Starter components. The photograph shows an overall view of a wye-delta starter for a 300-hp, 460V motor. The contents include: voltage-dropping resistors (arrow 1); a control transformer (arrow 2); fuses (arrow 3); control relays (arrow 4); line terminations (arrow 5); a time-delay relay (arrow 6); a line contactor (arrow 7); star and delta contactors (arrow 8); and dashpot overloads (arrow 9). The fact that this starter includes dropping resistors means that it is a "closed-transition" starter in which the wye (start) contactor opens before the delta (run) contactor closes. The dropping resistors allow the transition to take place without a momentary loss of power to the motor.

Some of the maintenance steps that should be taken are as follows.

Contactors should periodically be disassembled to check the condition of the contacts. The front of the contactor and the coil have been removed to reveal the movable contacts and the stationary contacts. Severely burned and pitted contacts should be replaced long before they are allowed to become heavily damaged. If the contacts are in poor condition, replacement kits are available. Trying to file or buff burned contacts is a waste of time and should not be attempted.

Terminating lug is of the single-barrel type used for one of the line-side feeder conductors. At least once a year, the connectors should be checked for proper torque. It is necessary to refer to manufacturer specifications to determine the proper torque. Too much torque is just as bad as not enough because damage to the connector can result. A torque wrench is a must.

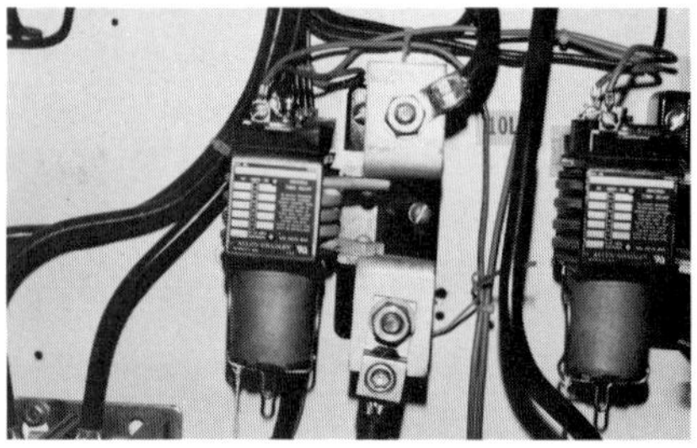

Dashpot overload relays are factory set for their particular application. They are generally maintenance-free items, but they can fail. Once a year, the dashpot should have its oil level checked by removing the reservoir. Inside the cup is an oil-fill line. If the level is below this line, some additional oil should be added. It is very important to use the type of oil recommended by the manufacturer because the operation of the dashpot overload relay is based on the viscosity of the oil.

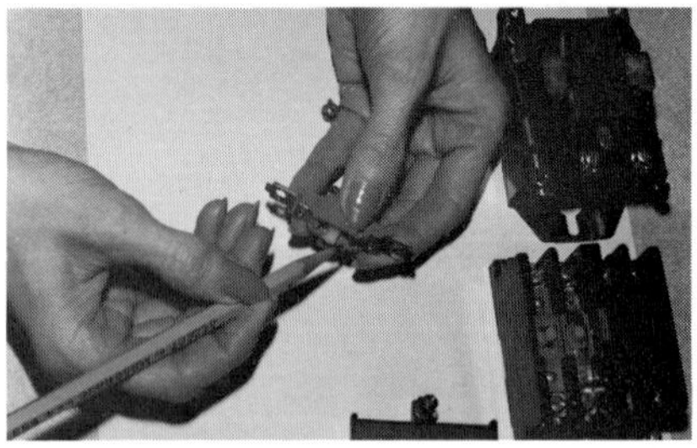

Control relays can be taken apart and the contacts inspected. A replacement set should be installed if the contacts appear to be burned. Making sure relay contacts are in good condition is a necessary part of a maintenance program. The proper operation of the wye-delta starter depends upon the condition of control components.

Wire terminations on the control relays and terminal strips should be checked at least once a year with a screwdriver to see that they are snug. Intermittent operation is one of the most frustrating parts of maintenance of starting equipment. Many times loose connections are to blame.

Voltage-dropping resistors generally are maintenance-free items for the life of the starter. However, it doesn't hurt to inspect them to make sure connections are tight, insulators are not cracked or broken, and that the resistance wire is not damaged in any way.

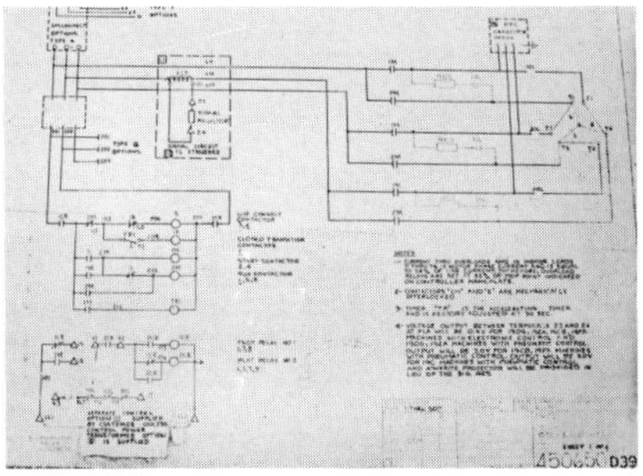

Documentation is probably the most neglected item in an industrial maintenance program. Making sure that the enclosure of the starter is clean and that an up-to-date schematic and wiring diagram of the wye-delta starter is stored in the cubicle are essential.

Dirt is the worst enemy of those servicing electromechanical equipment. Blowing off dirt and then vacuuming the enclosure once or twice a year is all that is needed. At the same time, the documents should be checked to see if any changes made in the wiring have been reflected in the drawings. Out-of-date drawings are the second worst enemy of electrical maintenance people.

Checking temperature of motors operating near to their full load should be done on a periodic basis. Elevated temperatures can lead to premature insulation failure.

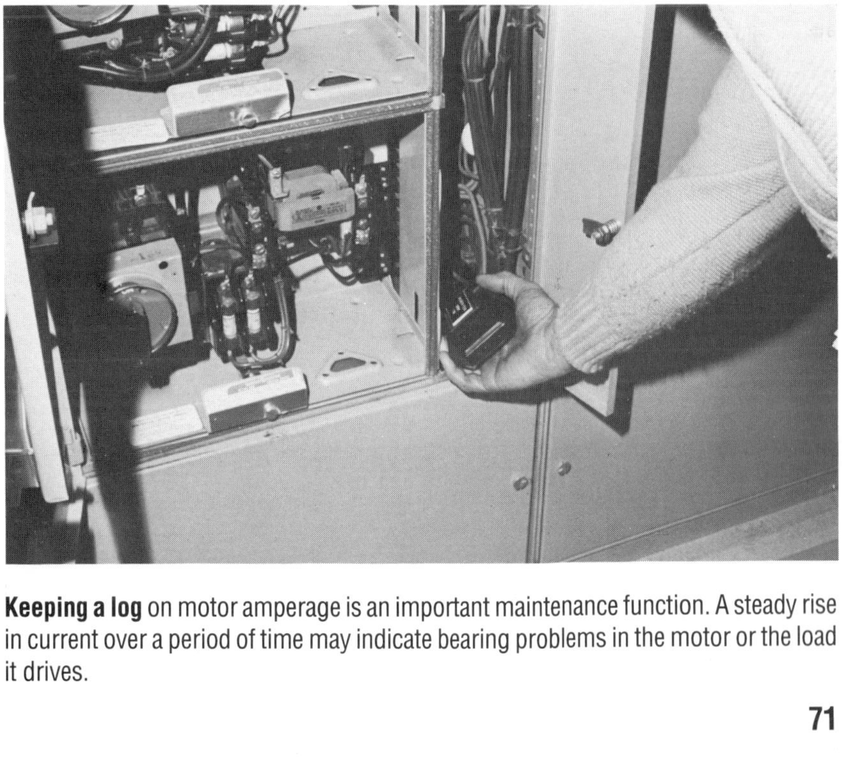

Keeping a log on motor amperage is an important maintenance function. A steady rise in current over a period of time may indicate bearing problems in the motor or the load it drives.

Preventive maintenance has been enhanced with the introduction of computer-based programs.

Temperature scanners are a recent major maintenance tool.

Contract maintenance firms often are called in to provide specialized services such as testing transformer oil.

Medium-voltage switchgear breakers and instruments require periodic testing and calibration to assure they will function when needed.

CASE HISTORIES

9

READING about the application of motors in different facilities is helpful to the designer, installer, and maintenance person in understanding the principles involved. These selected case histories deal with a variety of different motor types.

PAM MOTORS

St. Mary of Nazareth Hospital Center, Chicago, IL has slashed its electrical bill by replacing single-speed HVAC motors with pole-amplitude modulation (PAM) motors. The PAM motor is a single-winding, 2-speed induction motor that can be operated at a speed that will provide maximum effectiveness and efficiency.

On this project, 20 PAM motors ranging in size from 1.5 to 75 hp, were installed as replacements. As a result, the hospital is saving approximately $100,000 annually in reduced electric power costs. Nearly 75% of these savings can be attributed to just seven of the PAM motors ranging in size from 50 to 75 hp.

The 490-bed hospital is located in the West Town section of Chicago. Rising energy costs prompted hospital officals to evaluate a number of en-

PAM motors are two-speed, single winding motors, available in almost every standard size, such as this huge 5,000-hp example. The principal difference of the PAM and the standard squirrel-cage induction motor is in the stator winding coil connections that allow canceling the magnetic effect of some poles, thereby varying the speed. This allows the motor to operate as nonstandard ratios of speed; from as high as 10:2 to as low as 6:8.

Two of the three cranes used at VPA are shown undergoing testing. As increasing use is made of large container ships, modern automated systems of loading and unloading is required to avoid delays in port.

ergy-saving techniques. Grumman/ Butkus Associates, an energy management consultant firm, Evanston, IL, was retained to complete the study. Of sixteen energy-conservation measures identified, eleven were carried out.

One major project was to obtain better and more economical control of air distribution. It became apparent that the most cost-effective air-control technique was to reduce the operating level of certain fan systems during periods of reduced occupancy. Some key areas of the hospital, particularly those involving acute patient care, were found to require full ventilation 24 hrs a day. But many air-handling systems were serving areas that were occupied by 30 people during the day; but in some cases, only one person at night. The supply of full ventilation to these areas at all times was not necessary, and was a waste of energy and money.

The problem was finding the most affordable, energy-efficient means of reducing the level of operation for the motors. Deenergizing the motors completely was not considered because codes required that some level of air flow be maintained to all areas. Instead, it was decided to consider replacing the single-speed motors with two-speed versions, or to use variable-speed drives.

It was realized that there were disadvantages to using either technique for a retrofit operation. Variable-speed drives, for example, would have offered more versatility and energy savings, but were quite expensive, resulting in an excessively long payback period. Therefore, the retrofit was initially designed around standard, two-speed, two-winding motors. However, two-winding motors are larger than single-winding motors. Thus, their use would have required a change in mounting dimensions and extensive sheetmetal work to fit the larger motors. Also, two-speed motors operate at a 2-to-1 ratio between their high and low speeds. The low speed available from such a motor would not be sufficient to provide adequate air flow, even during periods of minimal occupancy. At this point that it was suggested that PAM motors be considered.

With the PAM motor a wider range of speed combinations with high/low ratios as close as 8-to-10 and 10-to-l2 is available. In this application, the speeds chosen were 1800/1200 rpm. Because of its single-winding design, the motors could fit the same frames previously occupied by the hospital's original single-speed motors.

After a thorough study, the final report estimated that the hospital would save nearly 2 million kWh of electricity annually, or about $72,000 (based on 1981 utility rates) in power cost savings, with PAM motors. The estimated cost for the project was calculated at $100,000 with a payback period of 18 months.

Since the new motors fit precisely onto the existing supports, average installation time was under 6 hrs. Time, materials, and labor costs were all less than what would have been required to install larger, two-winding motors.

DC DRIVES

Because of the increasingly large ships used to transport containers, Virginia International Terminals (VIT), the operations arm of the Virginia Port Authority (VPA), had to install new container cranes at VPA's Norfolk International Terminal (NIT), Norfolk, VA . To minimize the time that these larger ships must spend loading and unloading, port facilities use multiple cranes to handle a ship's cargo. However, there is a limit to the number of cranes that can be brought alongside a given ship. Thus, to further expedite the handling process, the productivity of the individual cranes had to be improved.

VIT and Liftech Consultants, Inc. of Oakland, CA, investigated the options and decided that a dual-hoist crane concept would best meet VPA's need. By modifying the basic design of the cranes and incorporating an operator-

Digital encoders (arrow) on a take-up reel are 40-kHz pulse tachometers that count the number of teeth that pass the location. This information is processed to give the digitally-controlled DC drives the exact positions of the trolleys, hoists, and intermediate platform.

supervised automatic positioning control system, the throughput capability of each new crane would be 50 containers per hr, which would be double the previous rate. This increased productivity would make the facility particularly attractive to carriers using these larger ships.

With the new arrangement, the positioning control for the cranes is made up of two programmable logic controllers (PLCs), digitally-controlled DC drives, and various proximity, load, and position sensors. A PLC-based local area network (LAN) provides a communications link between the control components, the PLCs, and a mainframe computer management information system (MIS).

Each crane is equipped with seven digitally controlled, fully regenerative, variable-speed DC drives that operate the 21 major motors on each crane. One drive operates two parallel, 320-hp, 250VDC ship-side hoist motors and eight series-parallel, 40-hp, 250VDC gantry travel motors, which are used to move the entire crane structure from berth to berth. Another drive operates two 450VDC, 175-hp motors individually for the boom hoist and trolley travel.

The dock-side DC control panel contains one DC drive to operate either the two parallel 320-hp, 450VDC dock hoist or one of the four 40-hp, 450VDC runway lift motors that are used to position the intermediate platform. Another drive operates either the 100-hp, 450VDC dock-side trolley travel motor or the second runway lift motor. The other two drives are dedicated to operating the two remaining runway lift motors.

An independently controlled ship-side operator's cab is run by the remaining DC drive. Located within the cab, this DC drive operates the two parallel, 5-hp, 450VDC motors used to position the operator's cab.

Motors used on these cranes have a "rated" and a "maximum" speed. The rated speed is the normal speed at which the motor should run carrying the maximum load. If the value of a load being handled is 50% of the maximum load, these drives will increase the motor's speed to about 200% of rated

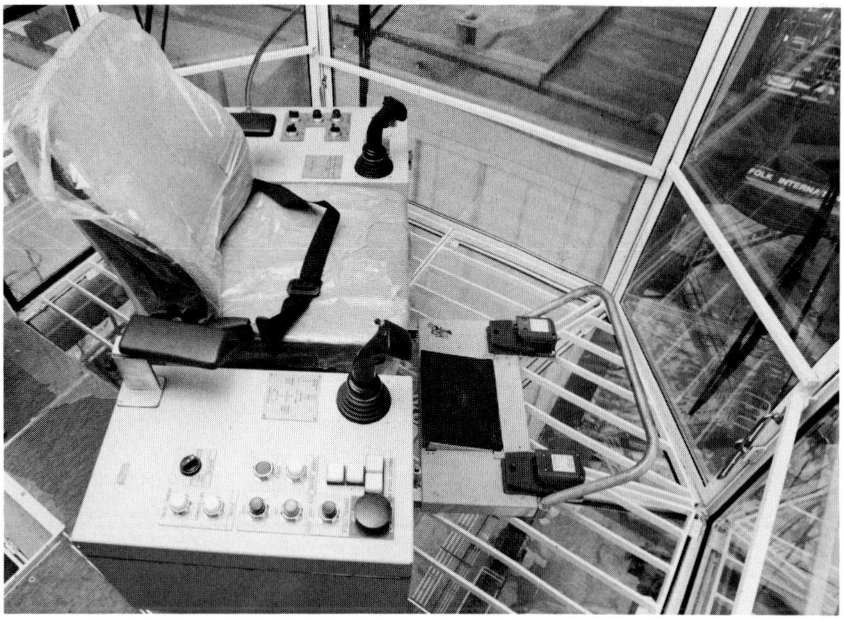

Control consoles on either side of the seat have pushbuttons that allow the operator to choose "offloading" or "onloading" cycles. The joysticks are used for manual operation during the "learning" process.

speed. The normal setting for the overspeed trip point in such applications is 10 to 15% above maximum speed. However, instead of using an absolute parameter, the drive's microprocessor permits calculation of the overspeed trip point based on the actual load and operating speed. So, at 50% load and 200% speed, the overspeed trip point is "set" by the microprocessor at a value that corresponds to the 200% rating plus 10 to 15%.

Digitally controlled units were selected over their analog-controlled counterparts for a variety of reasons. Motor operating parameters are set by inputting exact values directly into the drive's microprocessor instead of adjusting an assortment of potentiometers. In addition to allowing temporary changes or system fine-tuning, this eliminates the problems associated with potentiometer aging and temperature drift, thereby reducing maintenance requirements.

Memory devices used in the drive's microprocessor are erasable electrically programmable read-only memory (EEPROM) integrated circuit chips. EEPROMs can be programmed on one drive's microprocessor, removed without memory loss, and reinstalled in another drive. This allows preprogramming and verification of the individual motor's operating parameters at the factory or a workbench.

All position sensors for the hoists and trolleys are digital. Position encoding is done by incremental 40-kHz pulse tachometers. Mounted on the take-up reels for the hoists and trolleys in the main machinery room, these sensors count the number of teeth on the reel as it moves and provide the DC drives with that count. During automatic operation, these counts are the reference point from which the next position of the hoists/trolleys are calculated.

Proximity sensors are employed on the spreader and intermediate platform to tell the DC drives that the spreader has passed through a hatch or that a container is about to contact the intermediate platform. This information is used to generate speed changes in the drives.

Analog-type load sensors provide information on the actual load as well as misalignment information. If a filled container is heavier on one side than the other, the load sensors on the spreader indicate the eccentric nature of the load and allow the DC drives to compensate for container trim, list, and skew. This information is also fed to the mainframe computer to allow verification of the actual load against the listed load from the ship's manifest for

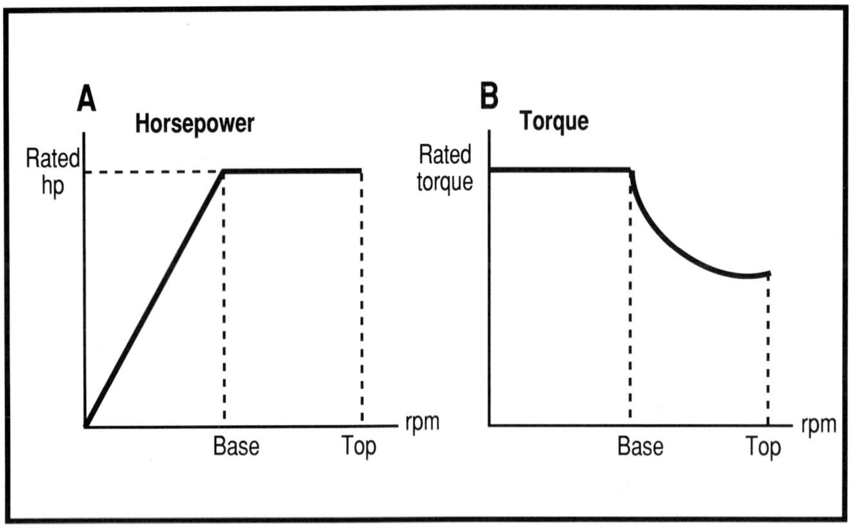

Fig. 9.1. Curves for DC motor showing hp vs rpm (A); and torque vs rpm (B).

instant identification of a container. This capability provides a record of the exact status and location of any container from the moment it is lifted.

Each crane is equipped with two PLCs. The PLC that controls ship-side operations is in the main machinery room. The other, in the auxiliary machinery room on the intermediate platform, controls the shore-side operations. A communications card, in the CPU rack, allows centralized control and monitoring of the DC drives and information exchange with the mainframe. Operations personnel utilize the LAN for status and diagnostic information, storing operating programs, and off-line modification or troubleshooting.

PLCs in this system also control the learn mode. In the learn mode, the PLC's processor "asks" the operator how far to travel in a specific direction. The operator must respond with movement in that direction only. When the desired position is reached, the processor checks the count from the digital encoders and commits it to memory. After positions for all movements are stored in memory, the processor and DC drives are capable of automatically directing the crane movements by performing elementary calculations to determine the crane's next position.

Although DC shunt-wound motors have many industrial applications, they are typically specified for crane applications because of their ability to change speeds smoothly over a wide range. They are used in cranes because they are more efficient for applications where frequent acceleration and deceleration of large inertia loads is required. In addition, these motors have high overload capabilities (as much as 300%), fast speed response, and excellent motor speed stability.

Shunt-wound DC motors were so named because at that time the armature and field circuits were connected in parallel to a single, constant potential power source. They now have separate connections for the armature and field circuits and use separate sources of DC voltage. A "commutating field" winding in series with the armature circuit is used to reduce armature reaction, which causes the motor's speed to increase with the load.

Speed control up to rated rpm is accomplished by increasing armature voltage while the field current is held constant. Horsepower delivered by the motor increases proportionally with the voltage (**Fig. 9.1A**). Speed control between rated speed and top speed is accomplished by reducing the field current, while the armature voltage and current remain at maximum. Although the actual torque value decreases, the hp rating remains constant because of the increase in speed (**Fig. 9.1B**).

HACR MOTORS

Replacement of chillers was part of the ongoing modernization of the electrical and mechanical systems within the Rockefeller Center complex in New York City. The original centrifugal compressor chiller equipment consisted of five units with ratings of 141, 205, 246, 512, and 688 tons, or a total capacity of 1792 tons. This array contained motors totaling 1950 hp.

In the first stage of this project, four of the chillers were removed from the basement chiller equipment room, with only the largest 688-ton chiller unit retained to temporarily provide supplemental cooling capacity until two new 1000-ton centrifugal liquid chillers could be brought on line. Each new unit has two 390-hp, hermetically sealed motors for a total of 780 hp. Thus, the combined total of 1560 hp for the two new chillers provides 2000 tons of cooling capacity, a considerable reduction in motor power requirements.

Circulating pump motor controller at the GE building uses a nonautomatic circuit breaker as a disconnect in series with 600A Class L current limiting fuses that provide short-circuit protection. The 250kcmil conductors from the contactor are carried overhead to the 150-hp, 208V motor.

An important feature of the new chiller unit's dual compressor design is its ability to operate efficiently at partial load, especially in the 300- to 700-ton output range. A microprocessor-based control system determines the proper compressor sequencing for all load conditions. As cooling needs change, guide vanes in the refrigerant vapor stream entering the compressors change position, initiated by a thermistor located in the exit chilled-water nozzle. This probe constantly transmits any variation in water temperature to the control unit. The microprocessor control, in turn, amplifies these signals and causes an actuator to adjust the guide vane position.

Temporary cable connections were made to a 390-hp hermetically-sealed motor that is part of a new 1000-ton centrifugal chiller. In the final installation, the conductors were run in flexible conduit.

Each of the four new 390-hp, 208V, 3-phase, centrifugal compressor motors are served by a wye-delta closed transition starter. For starting, a set of contactors connect the motor phase windings to form a wye circuit that produces 58% of normal line-voltage. Then for running, the motor phase windings are reconnected by the starter in a delta circuit to provide full line voltage. Thus, when initially energized, the motor draws about 250% of full-load current, much less than across-the-line starting (which is generally about six times normal full-load current). This avoids high inrush currents that can have an adverse effect on the power system by causing excessive voltage drop or transients.

Liquidtight flexible metal conduit provides flexibility to offset conductors from the overhead wireway to the circulating pump motor. The arrangement also limits the motor vibration that will be transmitted via the wireway to the starter.

Closed transition means that power is not removed during the change from reduced-voltage to full-voltage operation. A positive time-delay control circuit establishes the delay between the first stage of starting and the final full-voltage condition. In addition, the chiller's microprocessor control system places the load on the motor in a preprogrammed gradual manner, thereby minimizing starting torque and current inrush on startup.

For sealed (hermetic type) refrigeration compressor motors, the actual nameplate full-load running current of the motor is used in determining the current rating of the disconnecting means, the controller, branch circuit conductors, short-circuit, and motor overload protection.

MOTOR PROTECTION

A motor starter protection system saved the Buckingham Square Mall, Denver, CO. from failure of its centrifugal chillers that could have resulted in $30,000 to $100,000 worth of damage. A malfunction in electrical switchgear caused a fire and resulted in serious damage to the mall's main electrical system. The motor starter protection system (MSPS) protected the chiller motors not only during the initial fire, but also during a subsequent electrical system malfunction several days later.

Overload and undervoltage conditions can result in serious motor damage or total burnout and failure, while rapid out-of-phase reclosure of switchgear can impose torque stress on running motors and chiller impellers of up to 20 times what the components are designed to handle. This stress can cause damage to motor shafts, keys, and impellers.

Older chiller starters typically have an oil-filled dashpot overload protection device, which is subject to incorrect calibration and slow operation. The MSPS uses solid-state devices to provide faster operation and increased reliability with simple, infrequent calibration.

The MSPS detects problems and disconnects motors from the power source within six electrical cycles. In addition to distribution fault protection, it provides other features not found in typical phase and overload protection systems. Motor acceleration time, starter transition time, starter relay malfunctions, and inrush current level is all monitored.

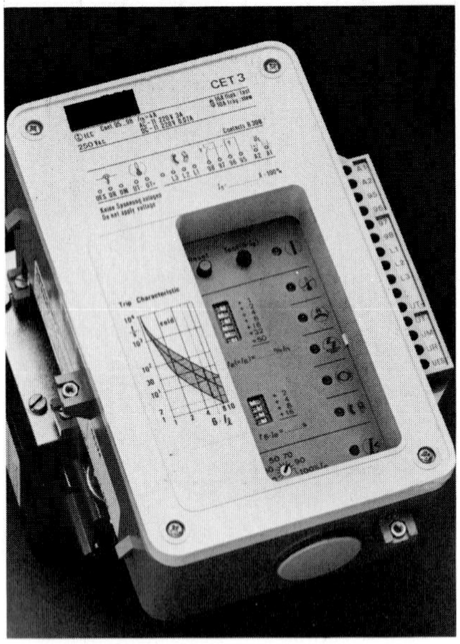

Motor protection equipment has advanced far beyond the thermal overload relay that guards against overloading. Monitoring devices can now supervise many characteristics of a critical motors and trigger an alarm or shutdown before the motor is damaged.

MAINTENANCE

At Port Elizabeth, NJ, six 140-ft-high gantry cranes must operate dependably; therefore, their medium-voltage AC and 500VDC systems require special attention. Thus, the primary consideration is the assurance of highest operational reliability, fast troubleshooting, and effective maintenance.

The cranes, each driven by eight 25-hp DC motors, roll on tracks near the edge of the dock. The DC power for these drive motors is obtained from a large motor-generator set installed in a machinery house on the main traveling boom of each crane. Festoon cables carrying 5kV conductors enter the house and are carried to a disconnect switch. Medium-voltage equipment in the 50 x 30-ft machinery house includes a 5kV motor starter that controls a 400-hp, 4160V induction motor that drives the crane DC power supply. A 320kW, 500VDC generator provides power for two 200-hp DC motors that drive the main lifting hoist and the eight 25-hp crane travel motors. A separate 65kW generator supplies DC power to a main boom travel 75-hp motor.

A maintenance program schedules work to be performed on a daily, monthly, and annual basis. Daily routines are quite simple and carried out by facility personnel and crane operators who check annunciator panels, meters, and gauges on vital equipment. Monthly and annual electrical maintenance procedures are more thorough and are done by outside contractor personnel.

Maintenance procedures cover the collector bus system, festoon cables, and DC equipment and systems. This program is important to ensure equipment and system reliability because of the hostile environment, which includes salt, dirt, carbon buildup, moisture, as well as wear caused by heavy equipment.

Two 5kV feeders from the substation to the dock area consist of lead-covered, crosslinked-polyethelene-insulated conductors. These conductors receive insulation-resistance tests, which are recorded and compared to earlier readings. Humidity and ambient temperature readings are also taken to provide correction for the resistance readings. All exterior conduits are painted to minimize corrosion.

Cabinets along the pier contain 5kV feeder terminations and 5kV contactors for operation of the 4160V collector bus system. The terminations are cleaned and inspected for any cracks or burn discolorations. The 5kV contacts are checked and cleaned. Because they are silver coated, they are never cleaned or smoothed with a file. The phase barriers are carefully checked because salt and carbon tracking has been found to be a problem. If tracking is detected, the barrier is replaced. The contacts of each contactor is checked with a spring scale. The pressure on all three contacts must meet the manufacturer's specifications and must be about the same value. If a significant difference is found on any one contact, all three springs are replaced.

Festoon cables and sheave cable guides are checked regularly by the operators for any signs of damage or wear. A more thorough inspection is carried out annually. Any worn or fatigued component is replaced.

DC generators, motors, and controls receive a thorough maintenance check each year. All rotating machine bearings are changed and checked with a solid-state stethoscope; and the temperature of the equipment is checked and recorded. All motors, generators, and circuits receive insulation-resis-

Maintenance is a must for critical electrical equipment driving the crane. All items are interdependent; failure of one will take to crane out of operation. The 400-hp, 4160VAC motor on the left drives the 320kW, 500VDC generator (center), which powers two 200-hp DC motors for the main lifting hoist, and the eight 25-hp DC crane travel motors. The 65kW DC generator on the right provides power for the main boom travel motor.

tance tests. Brushes are inspected for excessive wear, chatter or sparking, and commutators are checked for streaking, threading and other signs of possible problems. In addition, all equipment is cleaned and inspected and all terminations are checked for tightness.

SOLID-STATE STARTERS

Machining center. In a modern machining facility, frequent breakage of shear pins was occurring on a large, multitool, automatic milling machine. The machine was provided with a motor-driven chuck that holds the selected tool. The chuck motor was operated via traditional electromechanical controls that permitted excessive tightening of the chuck on the cutting tool. Because of this, when the tool had to be removed, the motor did not have sufficient torque to loosen the chuck. The shearpins, which protect against motor burnout, often failed. Excessive downtime was required to replace the pins.

A solid-state starter was substituted and used as a torque limiter providing adjustability to 70% of full load on tightening. The motor now runs to its stall point and shuts off. On reversal, the motor is given full current to break loose the tool for easy removal. In six months of usage, the machine shop has not sheared a single pin, which used to take 3 to 4 hrs to change every four to six weeks. With downtime estimated to cost $100 per hr plus parts and labor, the company estimates a savings of over $5000 per year.

Brick yard. At a South Carolina brick manufacturing facility, a pug mill grinds the clay used in making bricks. The mill is furnished with two 125-hp and one 200-hp AC induction motors. The three motors were originally controlled by reduced-voltage resistance starters; however, the mill officials were not happy with the performance and efficiency of these motor controllers. They decided to try a solid-state starter and quickly found the ideal combination of performance control that significantly increased efficiency of motor operation. After nine months of trouble-free operation, they purchased two more

Machine tools such as milling machines, grinders, and drill presses, are excellent applications for solid-state starters, particularly if machine startup or stopping requires precise control of motor inrush/stop current.

solid-state starters. The brick mill now conserves power and the starters provide a smooth, stepless start. The small cabinet size allows mounting the starter in close proximity to the mill.

Paper recycling center. This facility has a unit that chops and mixes paper into a fine pulp slurry. It was driven by a 300-hp motor controlled by an across-the-line starter that permitted a massive inrush current. This resulted in excessive torque that was damaging the drive couplings. A solid-state starter has significantly cut the inrush current and the accompanying torque, reducing component damage.

Water pumping station. At a booster station in Monrovia, CA, the city's water system contains two 100-hp pumps that supply water directly into the system. Solid-state starters with ramp-down features were specified to stop a water hammer effect that occurred when the booster pumps went on line and off line. These solid-state starters have eliminated the need for an expensive antisurge valve, which tries to eliminate or reduce the hammer effect mechanically. The starter does so more effectively, and is also more reliable and saves energy.

Cold rolling mill. In cold rolling metals, minor changes in rolling thickness can cause enormous changes in the loads on the drive train. It is possible for an inexperienced or unwary operator to damage the equipment. This resulted in the mill experiencing frequent damage to couplings, gearboxes, drive motors, and the planetary gear system. Downtime and repair costs were running high. Solid-state starters were selected to solve the problem because they could be adjusted to limit the load that the system will take and shut down the equipment if an operator tries to do more than the machine can handle.

MCCs

A major electrical modernization was recently completed at a large wastewater treatment plant in Tampa, FL. Among the new electrical items installed was a special multifunction integrated motor control center (MCC). The customized MCC houses four individual adjustable-frequency (AF) pump drives; a master flow/level control system with bubbler sensors; an annunciator; and a programmable logic controller (PLC) that coordinates all input/output signals to achieve desired pump-motor action.

Original pumping equipment included two 30-hp wound-rotor pump motors to pump during normal flow; a 75-hp squirrel-cage pump motor to pump during peak flow; and another 75-hp unit that served as a standby pump drive. The four pumps and their drive motors were retained in this update, except that the slip rings on the wound-rotor motors were shorted so that they could operate as induction motors and be controlled by AF drives.

The operating speed of each of the four pump motors now is controlled independently through its own AF drive. Flow-through of the system is regulated by running the proper pump or pumps as well as adjustment of pump-motor speed, resulting in the operation of any combination of the two 30- and 75-hp pumps as needed.

Decisions on which pump or pumps should run is made by the PLC in accordance with a schedule initially programmed into memory. When any pump reaches maximum flow demand, its drive motor is automatically switched to an across-the-line starter installed in the pump MCC.

All four AF drives operate in the same manner. Each unit controls the speed of its respective pump motor in accordance with its own level control signal, which is processed through the PLC. If the pumps do not run fast enough, the incoming sewage will accumulate in a level-monitoring tank, and the sewerage level will begin to rise until a correction is made to the flow. The level of this tank is a primary guide to flow requirements and is used to achieve the necessary speed change of the main pumps.

The sensor for this level measurement is a bubbler-type pneumatic system that measures the back-pressure of a stream of compressed air discharged near the bottom of the wet well. The pressure created is converted to a 4-20mA analog signal that is proportional to the level. This analog signal is then fed to the PLC where the analog signal is converted to DC signals so the data can be processed in its CPU.

The PLC output governs all AF controllers and their corresponding motor-pump speeds. Necessary sequencing, interlocking, signal processing, and alarm functions also are provided by the PLC.

The system annunciator provides visual and audible alarms. Major items monitored include: motor/pump failure, low flow, wet-well levels, AF drive failure, and valve operations. System status also annunciated include: manual or automatic operation; constant or variable speed; power supplies; and test functions.

By integrating the AF drives, starters, PLCs, alarms, etc. within the MCC, the amount of interwiring that was required to be done on the jobsite was greatly reduced. This also reduced the time needed to complete the project, and lowered the cost.

MCC equipment lineup consists of circuit breaker cubicles (arrow 1), adjustable-frequency drive controls (arrow 2), programmable controller (arrow 3), annunciator (arrow 4), and auxiliary motor controls (arrow 5).

QUESTIONS AND ANSWERS ON MOTOR CIRCUITS

10

OVER the years, EC&M has answered reader questions in its "Reader's Quiz" and "Code Forum" departments. Here is a selection of "hands-on" types of questions and answers that relate to the subjects of motors and their controllers.

Questions that are based on the NE Code are included here only to clarify some points, but the answers are current only for the 1993 issue of the code. Thus, before applying any of this information it will be necessary to check with the latest edition of the code to assure that no revisions have been made to the referenced sections.

MOTOR CALCULATIONS

Q1. How do you calculate the full-load current of a motor at various voltages?

A. The FLC of a 3-phase motor at various voltages can be determined with the equation:

$$I = \frac{hp \times 746}{1.73 \times \text{line volts} \times \text{efficiency} \times PF}$$

Notice that as the voltage increases, the FLC decreases when the hp remains constant. AC induction motors generally supply full hp at voltages varying by as much as +/- 10% of rated voltage.

Q2. How much full-load current (FLC) will a motor draw at other than rated voltage? For example, by a simple ratio calculation, will a 460V, 3-phase induction motor that draws 10A at 460V draw 10.4A when placed on a 480V line?

A. Using a voltage ratio to determine the difference in current drawn at different voltages is correct only for nonreactive loads. In the case of a motor, a reactive load is involved and the problem becomes much more complex. The ratio method will, however, give a fairly accurate ballpark figure if used correctly. In this question, the ratio has been inverted and a figure of 10.4A obtained rather than the correct answer of 9.6A.

An induction motor load is more accurately modeled as a constant-kVA load, in which case V x A = a constant value. Thus, at 460V

460V x 10A = 4600VA.

At 480V the current would be:

4600VA/480V =9.6A.

Actually, a 10% rise in terminal voltage will decrease the load current 7%, not 10% as derived using the constant-kVA equation. The reason is that as voltage rises, the magnetizing currents rise, offsetting the decrease in the current component associated with the motor shaft load.

Q3. What is a quick method of determining the full-load-current rating of motors that are not listed in NEC Table 430-150. Is there a rule-of-thumb method for finding the FLC?

A. One acceptable method for determining the FLC of a motor of a hp rating not listed in the table is as follows.

- From Table 430-150, select the listed motor of hp rating nearest to or the next higher hp than the unlisted motor.
- For the listed motor, divide the FLC in amperes by the hp rating to get the multiplier in amps-per-hp (A/hp).
- Multiply the hp of the unlisted motor by this multiplier to get the FLC in amps for the unlisted motor.

As an example, find the FLC of a 45 hp, 460V, 3-phase motor.

Table 430-150 does not list this size, but does list a 50 hp motor. It has a FLC of 65A. To obtain the multiplier:

65/50 = 1.3 A/hp.

The approximate FLC of the unknown motor is then calculated as:

FLC= 1.3 x 45 = 58.5A.

Q4. What is the effect on the torque of a 3-phase induction motor when the frequency is varied from 0 to 100% and from 100 to 150% with the voltage constant at all times; and when the frequency and the voltage are varied in accordance with the same values?

A. When a 3-phase induction motor is operated at different frequencies while the supply voltage is kept constant, the torque developed by the motor at a given speed will be approximately proportional to the square of the supply frequency. It is very important to note that the synchronous speed of the motor will be directly proportional to the supply frequency, and that the current drawn by the motor at a given speed will be approximately inversely proportional to the supply frequency.

A reduction in supply frequency reduces the magnetizing reactance. Therefore, if the supply voltage is kept constant, the magnetizing current (and consequently the magnetic flux in the air gap) will increase. This in turn increases the motor current. At very low frequencies, the magnetic circuit of the motor will operate far beyond saturation and the magnetizing current will become intolerably high.

When the supply voltage and the frequency are varied so that the ratio of the voltage-to-frequency remains constant, the flux density in the air gap remains almost constant. If resistance of the stator winding is neglected, the value of the maximum torque decreases somewhat with a decrease in frequency. (The synchronous speed is directly proportional to the supply frequency.) The slip for maximum torque varies inversely to the frequency, thus the maximum torque occurs at a lower speed when the frequency is decreased.

The equation for torque is:

T (in ft-lbs) = 5250 hp/rpm.

Thus, as frequency decreases, rpm decreases and torque increases. Using a 10-hp motor as an example, torque at 1800 rpm is 29.2 ft-lbs. Therefore, as frequency increases to 150% of normal in the same 10-hp motor, torque will decrease to 19.45 ft-lbs.

Q5. How does frequency effect the speed of a 3-phase AC motor?

A. The rpm of 3-phase induction motors varies directly with the frequency. Synchronous speed is calculated as:

S(rpm)= 120f/p

At 60 Hz, a 4-pole motor runs at 1800 rpm synchronous speed. At 1 Hz the speed is 30 rpm. A 2-pole motor runs at 3600 rpm synchronous, etc.

OTHER CALCULATIONS

Q1. Three single-conductors for a 3-phase, 60-hp, 230V motor are first routed inside a building in conduit, then underground outside the building in conduit, and then up a pole and overhead on a messenger to the motor. What size THWN copper conductors are required?

A. Calculations necessary to determine the proper wire size are as follows.

- Determine the required conductor ampacity. Table 430-150 gives the full load current of a 60-hp, 230V motor as 154A. Sec. 430-22(a) requires the conductors supplying a single motor to have an ampacity no less than 125% of the motor full-load rating. The required conductor ampacity:

  I= 1.25 x 154 = 192.5A.
- Size conductors for inside the building in conduit from Table 310-16. No. 3/0 THWN, rated 200A, is required.
- Size conductors for underground in conduit from Table B-310-7, based on one conduit with three conductors. No. 2/0 THWN, rated 200A, is required.
- Size overhead conductors, supported by a messenger, from Table B-310-2. No. 2/0, rated 212A, is required.

If a single set of conductors is used for the entire run, they must meet the most severe requirements, and therefore must be size No. 3/0. If the underground and overhead runs are long, this might not be economical. It is possible to splice from the No. 3/0 conductors in the building to No. 2/0 conductors for the underground run, using an accessible, approved underground splice box. Where the conductors come out of the ground and go up the pole in conduit, Table 310-16 again applies, so another splice box would be required to go back to No. 3/0 conductors. At the top of the pole, if the messenger-supported portion is long enough, it might then be economical to again splice to No. 2/0 conductors. Whether these splices from No. 3/0 to No. 2/0 conductors are made depends on the length of the runs, and the cost of the splices. This must be compared with the cost of No. 3/0 cable for the entire run, and the potential reduced reliability due to the splices.

Q2. In sewage treatment plants, lift pumps play important roles. They run when liquid levels reach specified levels, and continue to run for long periods of time. The motors are rated for continuous duty. When sizing the conductors for one of these motors, does the requirement of Sec. 430-22 that branch circuit conductors have an ampacity of not less than 125% of the motor full-load current rating apply, or must the conductors be sized by Exception No. 1, which refers to Table 430-22(a) Exception?

A. Article 100, under "Duty" has definitions of "continuous," "intermittent," "periodic," "short-time," and "varying" duty, and a fine-print note (FPN) refers to Table 430-22(a) Exception for illustrations of various types. The definition of continuous duty best covers the lift pumps described in the question.

Conductors with an ampacity of 125% of the motor full-load current rating are probably adequate. Sec. 430-22(a) Exception No. 1 and the accompanying table require larger conductors, 140 or 200% for continuous-rated motors, depending on the duty cycle. This requirement recognizes that motors draw large inrush currents with each start, and that if they start and stop frequently, these large currents might overheat the conductors supplying the motor unless the conductors are larger than for continuous operation.

If the duty cycle for the motors meets the definition of intermittent, short-time, periodic, or varying duty, the conductor sizes must be in accordance with the table. If the definition of continuous duty applies, the conductors must be sized at 125% of the motor full-load current rating.

Q3. Four 125-hp, 460V, 3-phase, 60-Hz electric motors drive centrifugal machines in a sugar processing plant. These motors are open-ventilated. Because the area where these motors are installed is normally humid due to the presence of steam drain lines and sugar vapor, the motor windings require cleaning, usually every three months. It was decided to modify the system by providing hoods to the motors to supply filtered air from outside the plant by means of a blower. What cubic ft per min (cfm) of air will be required to maintain proper operation if the average air temperature is 85°F?

A. The 125-hp motor probably has a full-load efficiency of about 90%. Therefore, the amount of power it will consume will be:

$$\frac{125 \text{ hp} \times (746\text{W per hp})}{90\%} = 103{,}611\text{W}$$

Of this power, 90% is transferred as mechanical energy to the driven machinery. The remainder is reduced to heat directly at the motor. Converting: 103,611W x 10% x 3.41 Btu/Wh = 35,331 Btu/hr.

Assuming the motor is designed to operate in an ambient temperature not to exceed 104°F (40C), the 85°F ventilation air can be permitted to rise 19°F while absorbing the heat of the motor.

The quantity of air required will be:

$$\frac{35{,}331 \text{ Btu/hr}}{(19 \times 1.08 \text{ Btu/hr})} = 1722 \text{ cfm per motor}$$

The factor 1.08 is derived from multiplying the specific heat of air (0.24 Btu/lb-°F), the density of air (0.075 lb/cu ft), and 60 min/hr.

If the motor in question is not fully loaded it may be that the quantity of ventilation can be reduced. However, it must be kept in mind that operating at less than full load decreases the efficiency, which will increase the proportion of heat produced at the motor.

Another way of looking at the problem is as follows. ASHRAE Fundamentals (Section 26) rate the heat gain from a 125-hp motor as 35,300 Btu/hr. If 85°F air is blown on each individual motor, and not more than 100°F air leaves after ventilation (40°C = 104°F), then 2179 cfm are required for each motor. This is obtained as follows:

$$Q \text{ in Btu/hr} = 1.08 \times \text{cfm} \times (T^{out}-T^{in})$$
$$= 1.08 \times 2179 \times (100-85)$$
$$= 35{,}300.$$

If the actual air properties are known, they should be used. Q is equal to hp of the motor (125 hp). Only the motor windings and bearing energy losses need be considered in computing the motor heat output since energy output is not in the ventilation air stream.

MAINTENANCE HINTS

Q1. What is the most effective way to locate the brush neutral position on a DC shunt motor?

A. The method most commonly used is the "kick" method. With all the brushes raised from the commutator and the machine not turning, the shunt field is excited to about one half its normal strength and the field current is suddenly broken. Voltage will be induced in the armature by transformer action. These induced voltages in conductors located at equal distances to the right and left of the main pole centers will be equal in magnitude and opposite in direction.

If the terminals of a low-reading voltmeter (5V) are connected to two commutator bars on the opposite side of the main pole and exactly half way between the center lines of two main poles, the voltmeter will show no deflection when the field current is broken. The spacing of these commutator bars is, therefore, the correct distance between brushes on adjacent brush arms. A digital voltmeter is completely worthless for making this test.

The most practical method of making this check is to make two pilot brushes of wood or fiber to fit the regular brushholder. Each brush should have in its center a piece of copper fitted for line contact with the commutator, and with a lead for connection to the voltmeter. By fitting two such pilot brushes in the holders of adjacent brush arms, the brush rigging may then be shifted slightly forward or backward, as necessary, until breaking the field current produces no deflection on the voltmeter. By noting the position at which no deflection is obtained for each pair of brush arms, the average of the positions of neutral obtained gives the correct running location for the brushes.

A quick and convenient method of locating neutral on a DC motor having shunt fields is to check the speed of the motor in either direction with the same impressed line voltage. The position of the brushes that produces the same speed in either direction under the same voltage conditions, is the correct neutral.

Another effective technique is to mark one armature coil slot with chalk and trace its leads to the commutator. Turn the armature in the motor so that the marked slot is under the interpole. With the armature in this position, move the brush holder so that one brush is over the commutator bars connected to the coil. Lock the brush holder in this position. Run the motor for a while with the brushes in this position, then shift the brushes back and forth slowly and notice if the motor runs more quietly or without any sparking. The location of the brushes one bar away from the determined position may cause better operation; if so, leave the brushes in the new position.

Q2. Preventive maintenance (PM) for large synchronous motors calls for the polarity of the slip rings to be changed every six months. I was told that this reduces wear on the slip rings. Is it really necessary to reduce wear on the slip rings?

A. Viable PM programs are refined as they go along. If anything needs more or less attention, it is so noted and the PM program schedule is adjusted. Thus the procedure probably is based on experience.

Slip rings do wear unevenly, and their polarities are changed periodically to extend their service lives. Cost-conscious facilities will include this operation in the regular PM program, as reversing the polarity of slip rings does save early replacement. Is it really necessary to reduce wear on the slip rings? Yes, if you want to control operating costs. No, if you mean will the motors fail to run after more than six months without the change. However, this kind of reasoning is the basis for future problems. It is always best to follow procedures that will avoid problems.

Q3. We started an insulation resistance (IR) testing program at a steel mill. No previous records existed. The temperature-corrected 1-min IR values on several DC machine armatures (up to 3000kW) are acceptable, but most of the polarization index (PI) values are poor with many between 0.86 and 1.5. The machines are rated 700VDC; the test voltage was 500VDC. The fields pose similar problems. In-place cleaning has not proven to be a satisfactory solution to raising PIs. Shop cleaning and baking improve PIs, but when returned to service (cold roll application), PI values fall to 1.2 (from 2.5) within one month. Motor enclosures are equipped with filters that are changed regularly. We would like to meet IEEE 43 guidelines of at least 1.7 megohm IR and a minimum PI of 1.5. What type of PI guidelines and/or solutions are recommended?

A. The polarization index is a ratio of insulation-resistance values recorded during a dielectric absorption test. A DC test voltage is applied to the equipment insulation and a current reading taken. The 10-min reading is ratioed to the 1-min reading and the quotient is the PI. The higher the PI, the better the dielectric strength of the insulation. After 10 min of test, the insulation resistance should have increased from the 1-min value because system charging currents will have had time to diminish. Where this is not the case, there is reason to believe insulation failure is imminent since leakage currents are persisting. A PI less than 1 is considered poor, while values between 1 and 2 are marginal, values between 2 and 4 are good, and above 4 the insulation system is excellent.

Many factors including moisture, ambient temperature, dirt accumulation, and motor winding temperature, influence the resultant test values. A trend log of the insulation testing should be kept so that conclusions can be made on the basis of historical test data.

If the DC machines having a PI less than 1 can be taken out of service without affecting operations, such action is recommended. For those machines having a PI greater than 1, what is recommend is establishing the trend log of armature and field insulation resistance and postponing rebuilding until meaningful trend log information can be gathered.

If monthly or quarterly temperature-corrected IR value is not showing a decline for an individual machine, then it is probably all right. Also, the older Class A insulation has a low temperature coefficient, which means that the IR readings vary more widely with temperature than Class B windings. One final point: when taking IR readings, the relative humidity should be recorded, as the winding can absorb moisture on a humid day and change the apparent IR value.

Another measure is to take the 30- to 60-sec IR readings to determine the dielectric absorption ratio (AR). This can also give an indication of possible faulty windings. An AR of less than 1.0 indicates a possible damaged wiring and an AR of 1.0 to 1.25 is questionable.

DISCONNECTS

Q1. A motor controller is installed within sight of the motor it controls, and the required disconnect for the motor and controller is within sight of the controller. However, the disconnect is not within sight of the motor and is not capable of being locked in the open position. Does this installation meet with the requirements of the NEC? If so, why does the code require the controller, rather than the motor disconnect, to be within sight of the motor?

A. The disconnect should be in sight of the motor and driven machinery. Sec. 430-102(b) states that a disconnecting means must be located in sight from the motor location and the driven machinery location. An exception is made where the disconnecting means provided in accordance with Section 430-102(a) is capable of being locked in the open position.

Since the disconnect in question is neither within sight of the motor and its driven machinery, nor capable of being locked in the open position, the installation does not meet NEC requirements.

Q2. It is the practice at many large industrial plants to supply 120VAC control power for 480VAC motor control centers (MCCs) from lighting and appliance branch-circuit panelboards located throughout the facility. Because each motor circuit in the MCC receives control power from a circuit of the panelboard, it is difficult to determine which breaker in which panelboard must be deenergized to disconnect control power to a given motor. Even when this circuit breaker can be identified, opening it disconnects control power to all the motors in the control center. Does this arrangement conflict with NEC requirements?

A. This arrangement is in conflict with NEC Sec. 430-74, which requires that the control circuits for motors must be so arranged that they will be disconnected from all sources of supply when the disconnecting means is in the open position. This section permits the use of two or more separate devices, one disconnecting the motor and controller from the sources of power and the other disconnecting the control circuits from their sources of power.

Sec. 430-103 requires that the disconnecting means disconnect the motor

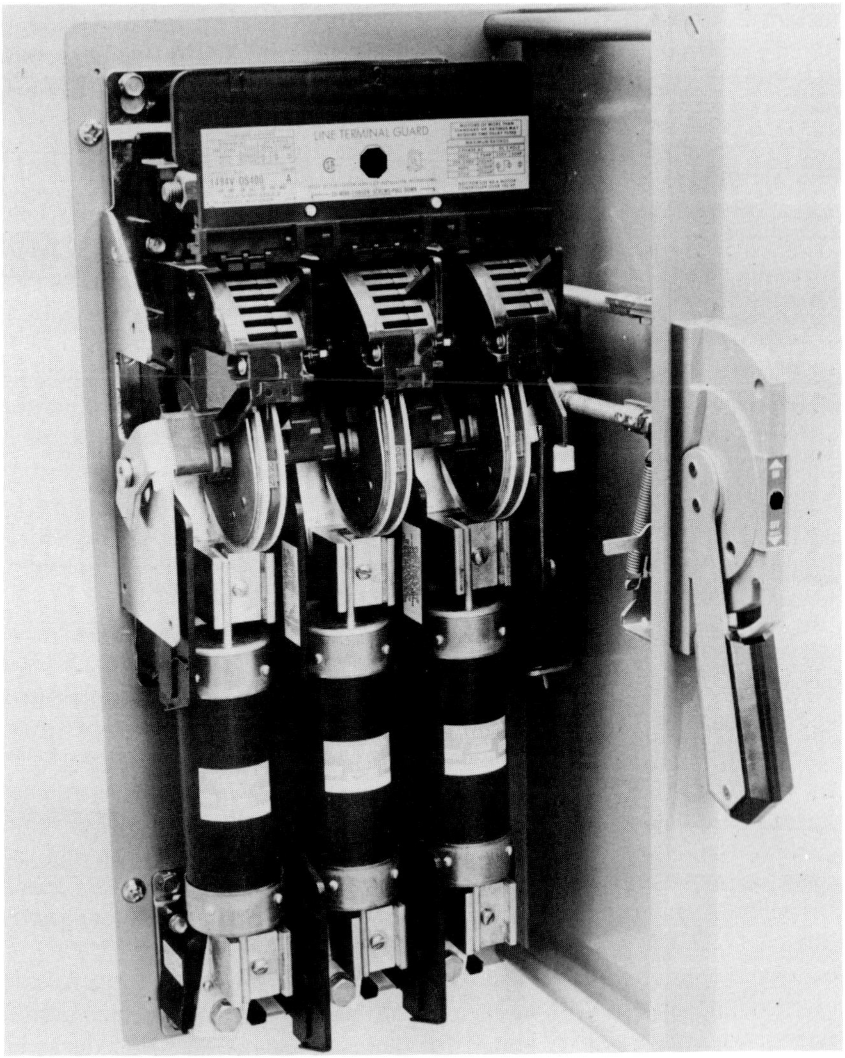

and controller from all ungrounded supply conductors. Sec. 430-112 requires each motor to be provided with a separate disconnecting means, and the exceptions do not apply to this case. Thus, the panelboard circuit breaker supplying the control circuit to the entire MCC cannot be used as the disconnecting means for control power to individual motors. The use of a microswitch or similar auxiliary device on the circuit breaker or other disconnecting means for each separate motor would be acceptable.

Q3. Heat-pumps that serve to provide air conditioning to individual hotel rooms are operate on 277V, 2-wire. The heat pumps are single units having two motors and are rated between 5.3 and 10A per unit. Can a toggle switch serve as the disconnecting means?

A. A toggle switch will meet this requirement where the rating of the toggle switch meets the provisions of Sec. 440-12 and, by reference, Sec. 430-109. Under Sec. 440-12(b), the ampere rating of the disconnecting means must be at least 115% of the total current drawn by the unit at full load, calculated as described in this section. This equals 6.1A for the 5.3A units, and 11.5A for the 10A units.

Also, Sec. 440-12 requires the horsepower rating of the disconnect to comply with Sec. 430-109, using the total current, including all motors and the resistance loads, as if it were the current of a single motor. For single-phase motors, the equivalent horsepower is obtained from Table 430-148. Since this table has no 277V column, the 230V equivalent current for the 277V known current must be derived. This is obtained by multiplying the 277V current by 277/230; the result is 6.4A for the 5.3A unit, and 12A for the 10A unit. Going to Table 430-148, this gives an equivalent of 3/4 hp for the smaller unit, and 2 hp for the larger one. These are the minimum permitted horsepower ratings for the respective disconnects.

Under Sec. 430-109, Exception No. 2, since the equivalent rating of the units is 2 hp or less at 300V or less, the disconnect may be a general-use switch not horsepower rated so long as the ampere rating of the switch is at least twice the full-load current, or 10.6 and 20A respectively.

This exception also permits the use on AC circuits of general-use snap switches suitable for use on AC (not AC-DC snap switches) if the switch has an ampere rating at least 125% of the full-load current, or 6.7 and 12.5A respectively.

If the toggle-type switch has at least these horsepower or ampere ratings, it may be used, subject to one additional consideration. The locked-rotor current also must be considered in determining the horsepower rating. This will not be a problem unless the locked-rotor current of any motor in the unit is more than six times the full-load current. If so, then the total locked-rotor current is used to determine the minimum horsepower rating of the disconnect as covered in Sec. 440-12(b). The minimum horsepower rating is the larger of the one determined by the running current and the one determined by the locked-rotor current.

Q4. A package-type central air-conditioning unit was installed on the roof of a commercial building. The mechanical inspector would not permit mounting a disconnect on the A/C unit. Therefore, the disconnect switch was located inside the building, remote from the roof. The disconnect switch can be locked in the open position. This time the electrical inspector rejected the job and said a disconnect would have to be installed near the A/C unit. Is the electrical inspector right?

A. Yes, the inspector is correct. The installation of the disconnect remote from the A/C unit is correct according to Article 430, Motor Circuits and Controllers. However, Article 440, Air Conditioning Equipment, applies in this case.

Sec. 440-14 requires that a disconnecting means must be located within sight from and readily accessible from air-conditioning or refrigeration equipment. There is no exception granted for a disconnect that can be locked in

the open position. However, Sec. 440-14 does state that the disconnecting means is permitted to be installed on or within the air-conditioning or refrigeration equipment.

PROTECTING MOTOR CIRCUITS

Q1. A 3/4-hp, 240V, single-phase motor was installed for a sprinkler system at a residence. No. 14 AWG THW copper conductors were used to feed this motor and the conductors were protected with a 20A CB. The local electrical inspector required that the wire size be changed to No. 12 AWG, based on the notes at the bottom of Table 310-16, which limit the value of overcurrent protection for a No. 14 AWG conductor to 15A.

The countering argument is that Sec. 240-3(f) permits motor circuit conductors to be protected in accordance with Article 430 and thus the rules of Article 430 take precedence and Sec. 430-52 applies in this case. Which interpretation is correct?

A. You are correct. The requirements of Article 430 supercede the notes at the bottom of Table 310-16, because it specifically states that "Unless otherwise specifically permitted elsewhere in this Code . . ." The application of motor and motor-control circuits is one of those situations where specific permission is otherwise given.

As called for in Sec. 430-22(a), the branch circuit conductors must have an ampacity not less than 125% of the motor full-load current. It can be seen in Table 430-148 that a ¾ hp single-phase, 230V induction motor, with no code letter, has a full-load current of 6.9A. To meet the requirement for a minimum conductor rating of 125%, a conductor with an 8.6A ampacity must be used. Sec. 210-19(c) also requires that the minimum acceptable conductor size for supplying branch circuits, for other than the household ranges and cooking appliances covered in Sec. 210-19(c), is No. 14 AWG copper. Therefore, the selection of the No. 14 AWG THW conductors is correct.

Branch circuit, short-circuit, and ground-fault protection for motors is covered in Sec. 430-52. The code requires that the motor overcurrent protection be calculated according to the values given in Table 430-152, which dictates the permissible multiplier to be applied depending on the type of overcurrent protection used. In this case, an inverse-time CB was selected to provide the required short-circuit and ground-fault protection for the motor circuit. From Table 430-152, a multiplier of 250% is applied to the full-load current given in Table 430-148. The multiplication shows that an inverse-time CB rated at:

$$2.5 \times 6.9A = 17.25A$$

is permissible for use as the overcurrent protective device in this application. Additionally, Exception No. 1 to Sec. 430-52 gives permission to use the next higher standard-rated protective device when the calculated value does not correspond to that of a standard-rated device and the next lower size is inadequate. According to Sec. 240-6, the next higher standard-rated device would be 20A. Therefore, in the application described, it would be permissible in accordance with the NEC to use No. 14 AWG THW copper conductors protected by a 20A inverse-time CB to make up this individual motor branch circuit.

Q2. A new 480V, 3-phase, 3-wire service was installed at an automobile mechanic's shop using three No. 4 AWG THHN conductors, which are terminated in a 100A, 600V fusible service disconnect. The service disconnect was fused at 35A because it only supplies a single feeder made up of three No. 10 AWG THHN conductors to feed a 30A, 600V fusible disconnect, fused at 20A. In turn, the disconnect feeds two machines: a 3-hp surface grinder and a 7 1/2-hp lathe.

Is it permissible to protect the No. 10 AWG THHN feeder conductors with a 35A-rated overcurrent device?

A. No. In selecting the size of a motor feeder overcurrent protective device, the NEC is concerned with establishing a maximum value for the fuse or CB. If a lower value of protection is suitable, it may be used.

To determine the rating or setting for overcurrent protection of a feeder supplying two or more motors, refer to Sec. 430-62(a). This section dictates that the required overcurrent protection must be rated or set at not more than the rating or setting of the largest branch circuit short-circuit protective device plus the sum of the full-load current(s) of all other motors.

Although no specific values for the individual branch circuit short-circuit protection were given in the initial question, the No. 10 AWG THHN conductors were terminated in a fusible disconnect, which was fused at 20A. It is only logical to assume that the individual branch circuit short-circuit protective devices are not rated in excess of this value.

Assuming that the largest branch circuit protective device is 20A, to comply with the requirements of Sec. 430-62(a), the full-load current of the other motors must be added. In this case, 20A + 4.8A (from Table 430-150 for a 460VAC, 3-hp induction motor, no code letter) = 24.8A. To comply with the requirement that the feeder short-circuit protective device be rated or set at no more than the rating or setting of the largest branch circuit short-circuit protective device plus the sum of the full-load current of all other motors, the rating or setting of the feeder overcurrent protection must be 20A. Had the value of the feeder overcurrent protection as calculated in accordance with Sec. 430-62(a) been greater than 30A, it would have been permissible to use a device rated or set at not more than that value.

Q3. Why do single-phase motors under nominal operating conditions occasionally burn out even though they have overload protectors. Could it be that the sizing of protectors to their motors is not being given the critical consideration it deserves, or perhaps the protector could be improved upon?

A. If the single-phase motor is a capacitor-start motor, a shorted capacitor could cause the starting winding to burn out before the overload protectors could act. Also, if the starting switch does not open the starting circuit when the motor reaches the appropriate speed, the starting winding will quickly burn out.

On the other hand, there are reasons other than overload protection that may be to blame. The motor environment may be at fault. Insufficient cooling/ventilation air will allow motors to heat

beyond their ratings and burn out. If motors have proper overload protection, but are heated from external sources to the burn-out temperature, the protective devices might well be innocent. Also, integral and fractional single-phase motors are often used in service where jogging is practiced more often than it is with larger 3-phase motors. Unless a motor is designed for jogging service, a few bumps in a short time will overheat it. The temperature of the motor will not be detected because it was not electrically overloaded. The end result will be a failed motor with overloads intact and functioning properly.

One certain protective device against burnout is an imbedded thermal switch. Motors can be purchased with this feature, or one can be installed when the motor is being rewound.

LUBRICATION

Q1. When lubricating motors, it is said that the condition, color, and smell of the old lubrication provides an indication of what may be happening in the bearing. For example, if the drained grease has turned into oil or its color has been darkened, it may indicate an excessively hot bearing. Are there any other visible "lubricant conditions" with corresponding possible problems, or other ideas that will help in the maintenance of motor bearings?

A. If a grease is permitted to remain in a bearing until it has turned to oil or has a burned odor, it is likely that the bearing will have been damaged. A statement in "A Guide to Better Bearing Lubrication," SKF Industries, Inc., says that the most critical factor when choosing a grease is the operating temperature of the running bearing.

One way to determine the temperature of the bearing cavity is to insert a thermocouple through the grease-filled hole and into the grease-filled bearing housing. The temperature obtained will be approximate, since the actual bearing temperature will be higher.

It is important to take into consideration the size and operating speed of the bearing. The larger the bearing and the higher the speed, the more frequently it should be relubricated (or a high-temperature type of grease must be used). When the temperature and operating conditions are known, it is easier to select the lubricant best suited for the job.

Also, it is important that the lubrication procedure, simple as it is, be performed properly. For example, if the motor is stopped, remove the drain plug (usually on the bottom of the bearing housing) and add new grease until all the old grease is removed (new grease begins to run out of the drain). Then run the motor for about 20 min before the drain plug is replaced. This will assure that any excess grease has been expelled from the bearing housing. Excess grease in the housing can cause bearing overheating.

Q2. How often is it necessary to grease a motor?

A. Instruction sheets that came with the motor should be followed exactly. They should be kept in the files of the person in charge of the maintenance department. The type of grease used must be known before a schedule of greasing can be determined.

Motors operating under normal conditions will have a greasing schedule of perhaps once a year. More severe conditions might require greasing every 6 months. Once a year, clean out and renew grease in ball and bearing housings. Use only a high-grade grease. It should be a little stiffer than vaseline and maintained over the operating range of temperatures induced in the operation of the motor. The melting point should not be below 150°C (302°F) and it should not show any signs of separation of oil or soap under operating or storage conditions.

Never overgrease or undergrease a motor bearing. If the motor does not have a drain plug, it should be taken apart at least once a year and have the bearings removed. The housing and the bearings should be cleaned with a good degreaser and regreased by hand, making sure that the bearing is filled. After reinserting the bearing in the housing, reassemble all parts except the outer plate or cap. Fill the housing one-third to one-half full. Any V-grooves found in the housing lip should be filled with a fibrous, high-temperature grease that will act as an additional protective seal against the entrance of dirt or foreign matter.

Q3. Is there really any difference between the various greases that are available for use in motor bearings?

A. There is, indeed, a difference and it is important for people who service motors to have a good knowledge of oils and greases.

Lime-base grease is used wherever moisture is prevalent. This grease, usually called cup grease, is composed of lard oil, horse fat, or mutton tallow combined with cottonseed or rapeseed oil, and pure mineral oil, with small quantities of stearic andoleic acids. Cup grease has a very wide application, and care should be used in the grade selected for a given application since there are several grades. When it is used in high-speed applications, the mineral-oil makeup should show a high viscosity test.

Soda-base grease makeup is generally the same as in the lime base, with the exception that caustic soda is substituted for hydrated lime. Since soda is soluble in water, these greases should not be used in applications where water is prevalent. Soda-based greases, also called sponge or fiber greases, are firmer than lime-based greases. They are applicable for lubrication of ball bearings.

Wool-yarn grease has a soda base, is much softer than sponge grease, and has the addition of approximately 10% wool waste. The wool waste adds to its cohesive properties.

Ordinary cup greases are quite frequently compounded with a calcium soap and have a melting point that is close to a motor's normal operating temperature. When the mineral oil separates out of the grease because of a high operating temperature, the residue remaining in the bearing housing is pure tallow and will clog alemite and zerke fittings. This also happens when a motor bearing is overgreased.